T0212293

Lung Sounds: An Advanced Signal Processing Perspective

Lung Sounds: An Advanced Signal Processing Perspective
Leontios J. Hadjileontiadis

ISBN : 978-3-031-00502-2   paperback

ISBN : 978-3-031-01630-1   ebook

DOI: 10.1007/978-3-031-01630-1

A Publication in the Springer series
SYNTHESIS LECTURES ON BIOMEDICAL ENGINEERING # 9

Lecture #9

Series Editor: John D. Enderle, University of Connecticut

Series ISSN

ISSN 1930-0328   print
ISSN 1930-0336   electronic

Lung Sounds: An Advanced Signal Processing Perspective

Leontios J. Hadjileontiadis
Aristotle University of Thessaloniki

SYNTHESIS LECTURES ON BIOMEDICAL ENGINEERING # 9

ABSTRACT

Lung sounds auscultation is often the first noninvasive resource for detection and discrimination of respiratory pathologies available to the physician through the use of the stethoscope. Hearing interpretation, though, was the only means of appreciation of the lung sounds diagnostic information for many decades. Nevertheless, in recent years, computerized auscultation combined with signal processing techniques has boosted the diagnostic capabilities of lung sounds. The latter were traditionally analyzed and characterized by morphological changes in the time domain using statistical measures, by spectral properties in the frequency domain using simple spectral analysis, or by nonstationary properties in a joint time–frequency domain using short-time Fourier transform. Advanced signal processing techniques, however, have emerged in the last decade, broadening the perspective in lung sounds analysis. The scope of this book is to present up-to-date signal processing techniques that have been applied to the area of lung sound analysis. It starts with a description of the nature of lung sounds and continues with the introduction of new domains in their representation, new denoising techniques, and concludes with some reflective implications, both from engineers' and physicians' perspective. Issues of nonstationarity, nonlinearity, non-Gaussianity, modeling, and classification of lung sounds are addressed with new methodologies, revealing a more realistic approach to their pragmatic nature. Advanced denoising techniques that effectively circumvent the noise presence (e.g., heart sound interference, background noise) in lung sound recordings are described, providing the physician with high-quality auscultative data. The book offers useful information both to engineers and physicians interested in bioacoustics, clearly demonstrating the current trends in lung sound analysis.

KEYWORDS

Lung sounds, advanced signal processing, nonstationarity, nonlinearity, non-Gaussianity, modeling, classification, denoising, heart sound cancellation

Contents

CHAPTER 1

The Nature of Lung Sound Signals

1.1 HISTORICAL OVERVIEW

ὅταν μὲν γὰρ ὁ τῶν πνευμάτων τῷ σώματι ταμίας πλεύμων μὴ καθαρὰς παρέχῃ τὰς διεξόδους ὑπὸ ῥευμάτων φραχθείς, ἔνθα μὲν οὐκ ἰόν, ἔνθα δὲ πλεῖον ἢ τὸ προσῆκον πνεῦμα εἰσιὸν τὰ μὲν οὐ τυγχάνοντα ἀναψυχῆς σήπει, τὰ δὲ τῶν φλεβῶν διαβιαζόμενον καὶ συνεπιστρέφον αὐτὰ τῆκόν τε τὸ σῶμα εἰς τὸ μέσον αὑτοῦ διάφραγμά τ᾽ ἴσχον ἐναπολαμβάνεται καὶ μυρία δὴ νοσήματα ἐκ τούτων ἀλγεινὰ μετὰ πλήθους ἱδρῶτος πολλάκις ἀπείργασται.

When the lung, which is the dispenser of the air to the body, is obstructed by rheums and its passages are not free, some of them not acting, while through others too much air enters, then the parts which are unrefreshed by air corrode, while in other parts the excess of air forcing its way through the veins distorts them and decomposing the body is enclosed in the midst of it and occupies the midriff thus numberless painful diseases are produced, accompanied by copious sweats.

Plato's *Timaeus*
On Physis: Diseases of the body (84d-e)
Translated by B. Jowett

From the time of ancient Greeks and their doctrine of medical experimentation until at least the 1950s, lung sounds (LS) were considered as the sounds originating from within the thorax and they were justified mainly on the basis of their acoustic impression. For example, writings of the Hippocratic school, in about 400 B.C., describe chest (lung) sounds as splashing, crackling, wheezing, and bubbling sounds emanating from the chest [1]. An important contribution to the qualitative appreciation of LS was the invention of the stethoscope by René Theophil Laennec in 1816. His gadget, which was originally made of wood, replaced the "ear-upon-chest" detection procedure enhancing the emitted LS [2].

Medicine was one of the first fields where conceptual tools of rationality and empiricism were combined with investigation techniques to make the human body an object of knowledge [3].

The use of the stethoscope played a tremendous role in medicine. It was not so much about the actual artifact as the technique that it crystallized, that is, mediated auscultation [4]. Thus, listening became important to the construction of medical knowledge and its application throughout the development of a technique and a technology to go with it, to such an extent that a doctor's hearing tool became the symbol of a profession, even as far back as the 1820s [5].

The development of the stethoscope and mediated auscultation coincided with the development of new theories of sense perception based on a "separation of the senses." To this end, seeing and hearing are to be understood as fundamentally and absolutely different modes of knowing the world, although neither form of knowledge is guaranteed as truth. This was the moment when empiricism collided with subjectivism [6]. If the sensorium was, before this moment, a type of complex whole, it then became an accumulation of parts. Thus, not only vision but hearing became its own, specific object of knowledge over the course of the 19th century, supplemented through technique and technology. From then on, audition became a key modality in perceiving states of patients' bodies.

Attempts for a quantitative approach date to 1930, but the first systematic, quantitative measurement of their characteristics (i.e., amplitude, pitch, duration, and timing in controls and in patients) is attributed to McKusick in 1953 [7]; the door to the acoustic studies in medicine was finally opened.

1.2   MAIN CHARACTERISTICS AND CATEGORIZATION

Sound, in general, consists of audible vibrations transmitted through an elastic solid or a liquid or gas, created by alternating regions of compression and rarefaction of the elastic medium. The density of the medium determines the ease, distance, and speed of sound transmission. The higher the density of the medium, the slower sound travels through it. Sound waves are characterized by the generic properties of waves, which are frequency, wavelength, period, amplitude, intensity, speed, direction, and polarization (for shear waves only).

In LS categorization, the principal characteristics, that is, frequency, intensity, duration, and quality (timbre/texture), are mainly considered. LS are divided into two main categories: normal (NLS) and abnormal (ALS). NLS are certain sounds heard over specific locations of the chest during breathing in healthy subjects. The character of the NLS and the location at which they are heard defines them. Hence, the category of the NLS includes the following types [8]:

- Tracheal LS are heard over the trachea having a high loudness.
- Vesicular LS (VLS) are heard over dependent portions of the chest, not in immediate proximity to the central airways.
- Bronchial LS (BLS) are heard in the immediate vicinity of central airways, but principally over the trachea and larynx.

- Bronchovesicular LS (BVLS), which refers to NLS with a character in between VLS and BLS, are heard at intermediate locations between the lung and the large airways.
- Normal crackles are inspiratory LS heard over the anterior or the posterior lung bases [9, 10].

ALS consist of LS of a BLS or BVLS nature that appear at typical locations (where VLS are the norm). ALS are categorized between continuous adventitious sounds (CAS) and discontinuous adventitious sounds (DAS) [11], and include the following types [12–14]:

- *Wheezes* (WZ) are musical CAS that occur mainly in expiration and invariably associated with airway obstruction, either focal or general, rhonchi, low-pitched sometimes musical CAS that occur predominantly in expiration, associated more with chronic bronchitis and bronchiectasis than with asthma.
- *Stridors* are musical CAS caused by a partial obstruction in a central airway, usually at or near the larynx.
- *Crackles* are discrete, explosive, nonmusical DAS further categorized as:

  - Fine crackles (FC)—high-pitched exclusively inspiratory events that tend to occur in mid-to-late inspiration, repeat in similar patterns over subsequent breaths, and have a quality similar to the sound made by strips of Velcro™ being slowly pulled apart; they result from the explosive reopening of small airways that had closed during the previous expiration.
  - Coarse crackles (CC)—low-pitched sound events found in early inspiration and occasionally in expiration as well, develop from fluid in small airways, are of a "popping" quality, and tend to be less reproducible than the FC from breath to breath.

- Squawks are short, inspiratory wheezes that usually appear in allergic alveolitis and interstitial fibrosis [15], predominantly initiated with a crackle, and caused by the explosive opening and fluttering of the unstable airway that causes the short wheeze.
- Friction rub are DAS localized to the area overlying the involved pleura and occur in inspiration and expiration when roughened pleural surfaces rub together, instead of gliding smoothly and silently.

The variety in the categorization of LS implies changes in the acoustic characteristics either of the source or the transmission path of the LS inside the lungs because of the effect of a certain pulmonary pathology. It is likely that the time and frequency domain characteristics of the LS signals reflect these anatomical changes [16]. In particular, the time domain pattern of the NLS

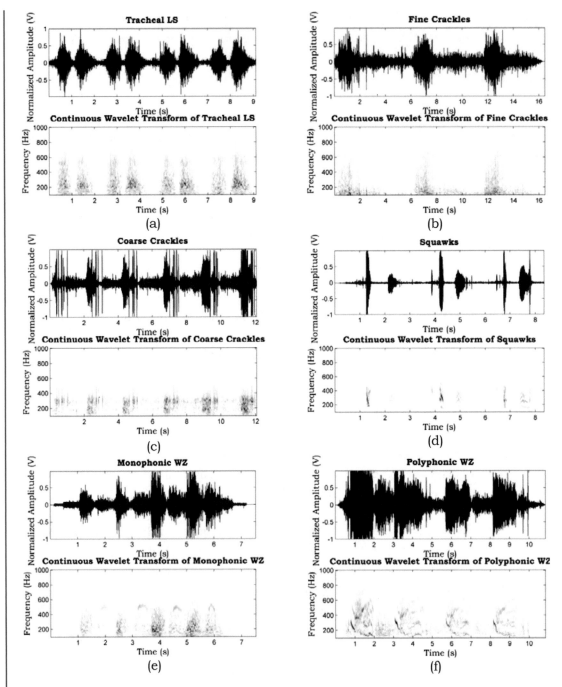

FIGURE 1.1: Typical examples of abnormal lung sounds (ALS). In all subfigures, the upper and lower parts illustrate the time- and time/frequency-domain representation of the ALS signal, respectively. (a) Tracheal LS, (b) fine crackles, (c) coarse crackles, (d) squawks, (e) monophonic wheezes, and (f) polyphonic wheezes.

resembles a noise pattern bound by an envelope, which is a function of the flow rate [16]. Tracheal sounds have higher intensity and a wider frequency band (0–2 kHz) than chest wall sounds (0–600 Hz) and contain more acoustic energy at higher frequencies [17]. The CAS time domain pattern is a periodic wave that may either be sinusoidal or a set of more complex, repetitive sound structures [16]. In the case of WZ, the power spectrum contains several dissonant-like peaks ("polyphonic" WZ) or a single peak ("monophonic" WZ), usually in the frequency band of 200–800 Hz, indicating bronchial obstruction [16]. Crackles have an explosive time-domain pattern, with a rapid onset and short duration [16]. It should be noted that this waveform may be an artifact of high-pass filtering [18]. Their time domain structural characteristics (i.e., a sharp, sudden deflection usually followed by a wave) provide a means for their categorization between FC and CC [19].

Some indicative examples of characteristic ALS are given in Figure 1.1. In this figure, the time domain and time/frequency domain representation (via the continuous wavelet transform; see Section 2.4) of ALS are depicted. These illustrations reveal the characteristics of ALS showing, for example, the explosive character of crackles superimposed on the vesicular sound (see Figure 1.1b, c) or the harmonic spread in the frequency domain when moving from monophonic wheezes to polyphonic ones (see Figure 1.1e, f).

For an extensive description and a variety of examples regarding LS structure and characteristics, the reader should refer to [16].

1.3 RECORDING ISSUES

1.3.1 Standards

LS auscultation is traditionally performed via the stethoscope. Listening to LS by means of a stethoscope involves several physical phenomena, such as vibrations of the chest wall that are converted into pressure variations of the air in the stethoscope, which are then transmitted to the eardrum. However, the stethoscope cannot be used in the quantitative analysis of bioacoustic signals (BAS), because it does not provide any means of signal recording. In addition, it presents a selective behavior in its frequency response instead of a flat one (at least in the area of interest, 40–4000 Hz).

One basic category of recording devices of LS signals refers to microphones. Whatever the type of microphone, it always has a diaphragm, like in the human ear, and the movement of the diaphragm is converted into an electric signal. Two major microphone approaches exist: the "kinematic" approach, which involves the direct recording of chest wall movement ("contact sensor"), and the 'acoustic' approach, which involves the recording of the movement of a diaphragm exposed to the pressure wave induced by the chest wall movement ("air-coupled sensor"). The chest wall movements are so weak that a free-field recording is not possible; it is essential to couple the diaphragm acoustically with the chest wall through a closed-air cavity. Whatever the approach, kinematic or acoustic, vibrations must be converted into electric signals using transduction principles of [20]: electromagnetic induction (movement of a coil in a magnetic field induces an electric current through the coil), condenser principle (changing the distance between the two plates of a charged capacitor induces a voltage fluctuation), and piezoelectric effect (bending of a crystal (rod, foil) for the induction of an electric charge on the surface).

The other category of recording devices includes piezoelectric accelerometers. A piezoelectric accelerometer applies the piezoelectric effect in such a way that the output voltage is proportional to the acceleration of the whole sensor. Early applications used heavyweight sensors with high sensitivity and good signal-to-noise ratio. The disadvantages, because of its heavy mass, are mechanical loading of the surface wall, difficulties with attachment, and a low resonance frequency (well within the band of interest). A piezoelectric accelerometer of very low mass (1 g) has been successfully applied in the past, yet it may be so fragile that routine clinical applications may be difficult [20].

Although both condenser microphones and piezoelectric contact sensors are displacement receivers, the waveforms that they deliver are different because of the coupling differences. Selection criteria of a device should also include size, average lifetime, and maintenance cost. Both sensors have several disadvantages, that is, piezoelectric sensors are very sensitive to movement artifacts (e.g., by the connecting wire, their characteristics depend on the static pressure against the body surface [21] and they are brittle), whereas condenser microphones need mounting elements that change the overall characteristics of the sound transduction [20].

1.3.2   Procedures and Considerations

The most common bandwidth for LS is from 60–100 Hz to 2 kHz when recorded on the chest, and from 60–100 Hz to 4 kHz when recorded over the trachea. In addition, recording of adventitious sounds on the chest requires a bandwidth from 60–100 Hz to 6 kHz [20]. Thus, sampling frequencies of 5–10 kHz are sufficient for the acquisition of all LS signals.

Electret microphones, such as Sony ECM 77, have a flat frequency response from 40 Hz up to 20 kHz in free space and are usually preferred in recording almost all types of LS. For the processing of LS, an analog system is required, which consists of a sensor, an amplifier, and filters that condition the signal before analog-to-digital (A/D) conversion (usually with a 12- or 16-bit resolution). A combination of low-pass filters (LPF) and high-pass filters (HPF) in cascade is usually applied. The purpose of using an HPF is to reduce the heart, muscle, and contact noises. The LPF is needed to eliminate aliasing. The amplifier is used to increase the amplitude of the captured signal so that the full A/D converter range can be optimally used, and sometimes to adjust the impedance of the sensor.

In recent years, some efforts of creating wireless sensors for LS recording have been proposed [22], based on Bluetooth® technology. The implementation of such concept could contribute to the continuous monitoring of LS and their archiving in a patient's database, thus realizing the concept

TABLE 1.1: Summary of recommendations for the case of LS signal acquisition with piezoelectric or condenser sensors [20]

SENSOR SPECIFICATION	
Frequency Response	Flat in the frequency range of the sound. Maximum deviation allowed, 6 dB
Dynamic range	>60 dB
Sensitivity	Must be independent of frequency, static pressure, and sound direction
Signal-to-noise ratio	>60 dB ($S = 1$ mV Pa^{-1})
Directional characteristics	Omnidirectional
COUPLING	
Piezoelectric contact	Immediate to the skin surface
Condenser air-coupled	Shape: conical; depth: 2.5–5 mm; diameter at skin: 10–25 mm; vented
FIXING METHODS	
Piezoelectric	Adhesive ring
Condenser	Either elastic belt or adhesive ring
NOISE AND INTERFERENCES	
Acoustic	Shielded microphones; protection from mechanical vibrations
Electromagnetic	Shielded twisted pair of coaxial cable
AMPLIFIER	
Frequency response	Constant gain and linear phase in the band of interest
Dynamic range	>60 dB
Noise	Less than that introduced by the sensor

TABLE 1.1: (*cont.*)	
HIGH-PASS FILTERING	
	Cutoff frequency: 60 Hz; roll-off: >18 dB octave^{-1}; phase as linear as possible; minimized ripple
LOW-PASS FILTERING	
	Cutoff frequency: above the higher frequency of the signal; roll-off: >24 dB octave^{-1}; minimized ripple

of "acoustic" file of each patient, something that is currently lacking from most clinical practice. Diffused distribution of many sensors over the patient's back and chest wall has also been proposed as a multichannel LS recording [23]. Based on these simultaneous multisensor recordings (e.g., ≥16 microphones), an acoustic mapping and imaging of the chest is feasible [24]. Although such an approach offers lower resolution compared to computerized tomography, X-ray, or functional magnetic resonance imaging techniques, it provides a new perspective in the exploitation of LS diagnostic information [25].

Table 1.1 tabulates recommendations [20] regarding the acquisition of LS signals.

1.4   TRENDS IN LS ANALYSIS

Although, in practice, LS interpretation is still based on the auscultative findings of the physician, modern computer-based analyses have been proposed as new means to extend LS capabilities by supplying information that is drawn not only directly from sound perception, but also from many other underlying characteristics that foster the diagnostic information of LS. Auscultation, however, remains a widespread technique, because it involves using the stethoscope. To this end, in most cases, the agreement of an automatic method with physician's interpretation is a basic criterion for its acceptance.

Through the years, LS analysis has branched out in various directions; the most popular include respiratory flow estimation [26, 27], heart sound cancellation [28–42], DAS detection and denoising [43–57], nonlinear analysis of LS [58–64], feature extraction, and classification [43, 52, 65–83].

In this book, focus is placed on two main categories of LS analysis from the perspective of advanced signal processing. In particular, the first one refers to the latest domains of LS representation that provide new insight on the structure and character of LS, forming new analysis tools, which can be of great assistance to the physician for a more efficient appreciation of the acquired LS. The second one includes up-to-date denoising techniques that take into account the nature of LS and provide enhanced signal quality, circumventing any noise effects in the interpretation of LS. The objective of presenting such methodologies is not only to reveal the importance of LS as indicators of respiratory health and disease, but also to shed light on the inherent characteristics of the advanced signal processing techniques that manage to adapt to the specific properties of LS, providing a new standpoint in the evaluation of respiratory acoustics.

·   ·   ·   ·

CHAPTER 2

New Domains in LS Presentation

2.1 OVERVIEW

Lung sound (LS) analysis has long existed between two main representations, that is, in time and frequency domain. Although LS time and spectral characteristics proved to be so far quite efficient in the interpretation of LS, they cannot always provide the full image of the diagnostic power of LS. To this end, some new analysis domains have been proposed to define some LS properties seen from different angles, capturing—perhaps, in a more pragmatic way—their behavior. These new representations and features of LS are examined in details in the subsequent sections.

2.2 HIGHER-ORDER STATISTICS (SPECTRA)

2.2.1 Epitomized Rationale

Questions about the Gaussianity, linearity, and minimum phase of the process under study often emerge within a biomedical signal processing framework. In the past, due to the lack of appropriate analytical tools, researches were forced to apply second-order (correlation-based) methods to just about every biomedical signal processing problem, ignoring those important distinguishing features. However, many real-world signals, such as almost all the biomedical ones, are actually non-Gaussian random processes. As a result, distinguishing features, such as non-Gaussianity, nonminimum phase, colored noise, nonlinearity, are important and must be accounted for in a signal processing context. To this end, in the recent decades, a class of tools has been developed to the point of practical application, known as higher-order statistics (spectra) (HOS), which have proven of potential value when dealing with non-Gaussian random processes [84].

HOS, also known as cumulants, are related to and may be expressed in terms of the moments of a random process. In general, the kth-order cumulant of a random process is defined in terms of the process's joint moments of orders up to k [84]. Just as the power spectrum (PS) (the Fourier transform of the autocorrelation) is a useful tool in the signal analysis, so too are the higher-order spectra, also known as polyspectra, which are the associated Fourier transforms of the cumulants. In PS estimation, the process under consideration is treated as a superposition of statistically uncorrelated harmonic components and the distribution of power among them is the estimation outcome.

As such, only linear mechanisms governing the process are investigated because of the suppression of the phase relations among the frequency components [85]. Consequently, the information contained in the PS suffices for the complete statistical description of a Gaussian process of known mean. Looking beyond the PS to obtain information regarding deviations from Gaussianness and presence of nonlinearities, the polyspectra should be used. The most often used ones are the third- and fourth-order spectra, also called *bispectrum* and *trispectrum*, respectively. From this perspective, the PS should be considered, in fact, as a member of the class of higher-order spectra, that is, as a second-order spectrum.

Cumulants not only reveal amplitude information about a process, but also reveal phase information. This is of great importance, because, as is well known, second-order statistics, that is, correlation, are phase-blind. A key characteristic, which differentiates cumulants from correlation, is that cumulants are blind to all types of Gaussian processes, whereas correlation is not. From a practical point of view, this means that when cumulant-based methods are applied to non-Gaussian (or, possibly, nonlinear) processes contaminated by additive Gaussian noise (even colored Gaussian one), they provide an analysis field of automatically improved signal-to-noise ratio (SNR).

Practically speaking, many biomedical signal processes are non-Gaussian, yet corrupted by measurement noise, which can often be realistically described as colored Gaussian process; hence, in these practical applications, the value of the HOS is apparent. Another very attractive property of cumulants is that they enable handlers to work with them as operators. On a practical level, this means that the cumulants of the sum of two statistically independent random processes equals the sum of the cumulants of the individual processes. Higher-order moments, by contrast, do not have that property; hence, they are much less convenient [84].

Based on the aforementioned HOS characteristics, placing LS in the HOS-based domain could reveal their true character, and hence, could further divulge their diagnostic value.

2.2.2 HOS Definitions: Higher-Order Statistics

Given a set of n real random variables $\{x_1, x_2, \ldots, x_n\}$, their *joint cumulants* of order $r = k_1 + k_2 + \cdots + k_n$, are defined as the coefficients in the Taylor expansion (provided it exists) of the second characteristic function $\tilde{\Psi}(\omega_1, \omega_2, \ldots, \omega_n) = \ln\left[E\left\{\exp(j(\omega_1 x_1 + \omega_2 x_2 + \cdots + \omega_n x_n))\right\}\right]$, about zero [86, 87], that is,

$$\mathrm{Cum}\left[x_1^{k_1}, x_2^{k_{21}}, \ldots, x_n^{k_n}\right] \equiv (-j)^r \frac{\partial^r \tilde{\Psi}(\omega_1, \omega_2, \ldots, \omega_n)}{\partial \omega_1^{k_1} \partial \omega_2^{k_2} \cdots \partial \omega_n^{k_n}}\bigg|_{\omega_1 = \omega_2 = \cdots = \omega_n = 0}. \qquad (2.1)$$

If $\{X(k)\}$, $k = 0, \pm1, \pm2,\ldots$ is a real stationary random process and its moments up to order n exist, then its nth-order cumulants, that is, its nth-order statistics, are given by the following equation:

$$c_n^x(\tau_1,\ldots,\tau_{n-1}) \equiv \mathrm{Cum}\left[X(k), X(k+\tau_1),\ldots,X(k+\tau_{n-1})\right]. \tag{2.2}$$

In the case where $\{X(k)\}$ is Gaussian, its third- and fourth-order cumulants, $c_3^x(\tau_1,\tau_2)$, $c_4^x(\tau_1,\tau_2,\tau_3)$, are identically zero. This can be easily deduced from the rewriting of (2.2) for a real zero-mean stationary random process $\{X(k)\}$ $k = 0, \pm1, \pm2,\ldots$, for $n = 3, 4$, in the form of

$$c_n^x(\tau_1,\ldots,\tau_{n-1}) = E\left\{X(\tau_1)\ldots X(\tau_{n-1})\right\} - E\left\{G(\tau_1)\ldots G(\tau_{n-1})\right\}, \tag{2.3}$$

where $\{G(k)\}$ is a Gaussian random process with the same second-order statistics as $\{X(k)\}$, and $E\{\cdot\}$ denotes the expectation value. What (2.3) clearly states is that cumulants not only display the amount of higher-order correlation, but also provide a measure of a distance of the random process from Gaussianity. Clearly, if $\{X(k)\}$ is Gaussian the cumulants are all zero; this is not only true for $n = 3$ and 4, but for all n.

For n stationary processes $\{X_i(k), i = 1, 2, \ldots, n\}$, the nth-order *cross-cumulant* sequence is defined as:

$$c_n^{x_1,\ldots,x_n}(\tau_1,\ldots,\tau_{n-1}) \equiv \mathrm{Cum}\left[X_1(k), X_2(k+\tau_1),\ldots,X_n(k+\tau_{n-1})\right]. \tag{2.4}$$

The relationship between fourth-order moments and cumulants for zero-mean signals [86] is given by

$$\mathrm{Cum}\left[x_1 x_2 x_3 x_4\right] = E\{x_1 x_2 x_3 x_4\} - E\{x_1 x_2\}E\{x_3 x_4\} - E\{x_1 x_3\}E\{x_2 x_4\} - E\{x_1 x_4\}E\{x_2 x_3\}. \tag{2.5}$$

For a real zero-mean stationary random process $\{X(k)\}$, $k = 0, \pm1, \pm2,\ldots$, by setting $\tau_1 = \tau_2 = \tau_3 = 0$ in (2.2), we obtain [86]:

$$\begin{aligned}
\gamma_2^x &= c_2^x(0) = E\{X^2(k)\} \text{ (variance)} \\
\gamma_3^x &= c_3^x(0,0) = E\{X^3(k)\} \text{ (skewness)} \\
\gamma_4^x &= c_4^x(0,0,0) = E\{X^4(k)\} - 3\left(\gamma_2^x\right)^2 \text{ (kurtosis)}.
\end{aligned} \tag{2.6}$$

Skewness shows the degree of asymmetry of a distribution. Kurtosis is a measure of whether the data are peaked or flat relative to a normal distribution. That is, datasets with high kurtosis tend to have a distinct peak near the mean, decline rather rapidly, and have heavy tails. Datasets with low kurtosis tend to have a flat top near the mean rather than a sharp peak; a uniform distribution would be the extreme case.

2.2.3   HOS Definitions: Higher-Order Spectra

Assuming that $c_n^x(\tau_1,\ldots,\tau_{n-1})$ is absolutely summable, the nth-order polyspectrum is defined as the $(n-1)$-dimensional discrete Fourier transform of the nth-order cumulant [84], that is,

$$C_n^x(\omega_1,\ldots,\omega_{n-1}) = \sum_{\tau_1=-\infty}^{\infty} \cdots \sum_{\tau_{n-1}=-\infty}^{\infty} c_n^x(\tau_1,\ldots,\tau_{n-1}) \cdot \exp\left[-j\sum_{i=1}^{n-1}\omega_i\tau_i\right]$$

$$|\omega_i| \le \pi, \; i = 1, 2, \ldots, n-1; \; |\omega_1 + \omega_2 + \cdots + \omega_{n-1}| \le \pi. \qquad (2.7)$$

In general, $c_n^x(\omega_1,\ldots,\omega_{n-1})$ is complex for $n > 2$ and thus it can be expressed as

$$C_n^x(\omega_1,\ldots,\omega_{n-1}) = |C_n^x(\omega_1,\ldots,\omega_{n-1})|\exp\left[j\psi_n^x(\omega_1,\ldots,\omega_{n-1})\right]. \qquad (2.8)$$

The nth-order polyspectrum is a 2π periodic function. The PS, bispectrum, and trispectrum are special cases of (2.7), that is,

- Power spectrum ($n = 2$):

$$C_2^x(\omega) = \sum_{\tau=-\infty}^{\infty} c_2^x(\tau) \cdot \exp\left[-j(\omega\tau)\right], \;\; |\omega| \le \pi \qquad (2.9)$$

- Bispectrum ($n = 3$):

$$C_3^x(\omega_1,\omega_2) = \sum_{\tau_1=-\infty}^{\infty} \sum_{\tau_2=-\infty}^{\infty} c_3^x(\tau_1,\tau_2) \cdot \exp\left[-j(\omega_1\tau_1 + \omega_2\tau_2)\right],$$

$$|\omega_1| \le \pi, \;\; |\omega_2| \le \pi, \;\; |\omega_1 + \omega_2| \le \pi \qquad (2.10)$$

- Trispectrum ($n = 4$):

$$C_4^x(\omega_1,\omega_2,\omega_3) = \sum_{\tau_1=-\infty}^{\infty} \sum_{\tau_2=-\infty}^{\infty} \sum_{\tau_3=-\infty}^{\infty} c_4^x(\tau_1,\tau_2,\tau_3) \cdot \exp\left[-j(\omega_1\tau_1 + \omega_2\tau_2 + \omega_3\tau_3)\right],$$

$$|\omega_1| \le \pi, \;\; |\omega_2| \le \pi, \;\; |\omega_3| \le \pi, \;\; |\omega_1 + \omega_2 + \omega_3| \le \pi \qquad (2.11)$$

The normalized bispectrum or *bicoherence* (Bic) is defined as [86]:

$$\mathrm{Bic}(\omega_1,\omega_2) = \frac{C_3^x(\omega_1,\omega_2)}{\left[C_2^x(\omega_1)\,C_2^x(\omega_2)\,C_2^x(\omega_1+\omega_2)\right]^{1/2}}, \qquad (2.12)$$

where $C_2^x(\omega_i)$, $i=1, 2$ is the PS defined in (2.9). From (2.12), it is apparent that bicoherence is a function that combines two completely different entities: the bispectrum, $C_3^x(\omega_1, \omega_2)$, and the PS, $C_2^x(\omega)$, of the process. The magnitude of bicoherence

$$\mathrm{BI}(\omega_1, \omega_2) = |\mathrm{Bic}(\omega_1, \omega_2)| \qquad (2.13)$$

is called the bicoherence index (BI) and its estimation is not bounded above by 1.0 [88]. Under the Gaussian assumption, the value of the bicoherency is zero. The $\mathrm{BI}(\omega_1, \omega_2)$ quantifies the presence of *quadratic phase coupling* (QPC) between any two frequency components in the process due to their nonlinear interactions. Two frequency components are said to be quadratically phase coupled (with $\mathrm{BI}(\omega_1, \omega_2)$ close to or greater than 1.0) when there exists a third frequency component whose frequency and phase are the sum of the frequencies and phases of the first two components [86]. *Cross-bicoherence* refers to two N-sample processes $\{X(k)\}$, $\{Y(k)\}$, and is defined through the cross-bispectrum, $C_3^{xxy}(\omega_1, \omega_2)$, as

$$\mathrm{CBic}(\omega_1, \omega_2) = \frac{C_3^{xxy}(\omega_1, \omega_2)}{\left[C_2^x(\omega_1)C_2^x(\omega_2)C_2^y(\omega_1 + \omega_2)\right]^{1/2}}. \qquad (2.14)$$

The magnitude of cross-bicoherence

$$\mathrm{CBI}(\omega_1, \omega_2) = |\mathrm{CBic}(\omega_1, \omega_2)| \qquad (2.15)$$

is referred as *cross-bicoherence index* (CBI), and quantifies the presence of QPC between any two frequency components in the two processes due to their nonlinear interactions.

For a detailed description of HOS properties, symmetries, and estimators, the reader is referred to [86] and references contained therein.

2.2.4   The Parametric Approach

Moving on to a parametric estimation of bispectrum, we consider a real, zero-mean, and stable ARMA(p, q) process $\{X(k)\}$ given by

$$\sum_{i=0}^{p} a(i)X(k-i) = \sum_{j=0}^{q} b(j)W(k-j), \qquad (2.16)$$

where $\{W(k)\}$ are independent and identically distributed (i.i.d.) random variables, independent from $\{W(m)\}$ for $m < k$, with zero mean and

$$E\{W(k)W(k+m)\} = Q\delta(m) \qquad (2.17)$$

$$E\{W(k)W(k+m)W(k+n)\} = \beta\delta(m,n)$$

(2.18)

$\beta\delta(m, n)$ denotes Kronecker's delta function, $Q = E\{W^2(k)\}$ and $\beta = E\{W^3(k)\}$, that is, $\{W(k)\}$ is third-order white. Note that both $\{W(k)\}$ and $\{X(k)\}$ are non-Gaussian. The bispectrum of the system input and output are related by

$$B(\omega_1, \omega_2) = \beta H(\omega_1)H(\omega_2)H^*(\omega_1 + \omega_2),$$

(2.19)

where $H(\omega)$ is the frequency response function of the autoregressive moving average (ARMA) system given by

$$H(\omega) = \frac{\sum\limits_{\ell=0}^{q} b(\ell)e^{-j\omega\ell}}{1 + \sum\limits_{\ell=1}^{p} a(\ell)e^{-j\omega\ell}}.$$

(2.20)

2.2.5   LS Quadratic Phase Coupling Detection

QPC refers to peaks at harmonically related positions in the PS. Because the existence of phase-coupling denotes the existence of nonlinearity, which, in turn, in many cases, is related with the existence of pathology, detection of QPC in LS signals using the BI could possibly provide an indicator of anatomical changes of lungs due to pathology.

A nonlinear analysis of musical LS based on BI (see (2.13)) is presented in [59]. The third-order statistics (TOS) and spectra were used both for the detection of QPC between distinct frequency components of musical LS and as a measure of their nonlinear interaction. This harmonic analysis was conducted on preclassified signals (wheezes, ronchi, stridors) selected from international teaching tapes. The derived results have shown a high degree of deviation from Gaussianity along with quadratic self-coupling ($f_2 = 2f_1$; $\phi_2 = 2\phi_1$) within the distinct frequencies of the corresponding PS of musical LS. This is shown in Figure 2.1, where an example of the recorded LS (inspiratory stridor recorded from an infant with croup [59]), the estimated PS, and bispectrum are depicted in Figure 2.1a, b, and c, respectively. From the three distinct peaks seen in the PS (i.e., 410.2, 820.4, 1230.6 Hz), a self-coupling is revealed in the bispectrum domain denoted by the clear peak at (410.2, 410.2) Hz.

Differences in the degree of QPC as well as the nonlinearity and the non-Gaussianity of the processed LS signals could establish a new field in characterization and feature extraction of the pulmonary pathologies associated with musical LS, such as asthma and chronic obstructive pulmonary disease (COPD) [59].

As reported in [52, 58], in the case of monophonic wheezes, the frequency pair of QPC belongs to the low frequency band. The fundamental frequency is revealed in the inspiratory phase

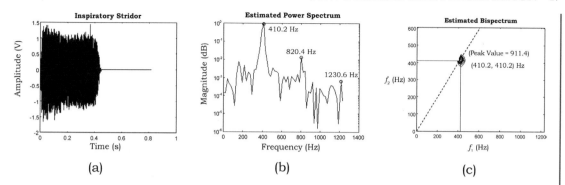

FIGURE 2.1: An example of QPC seen in the case of a recorded inspiratory stridor depicted in (a) from an infant with croup [59], (b) the corresponding PS, (c) the corresponding bispectrum estimated within the principal region in the bifrequency domain. The distinct peak in the bispectrum denotes existence of the QPC and strong deviation from the Gaussian assumption.

of a breath cycle, whereas its first harmonic dominates in the expiratory phase. This fact leads to the conclusion that monophonic wheeze consists of a single note or a single tonality, established by the octaves of the fundamental frequency, due to a nonlinear mechanism. In the case of polyphonic wheezes, the frequency pair of QPC belongs to a higher frequency band than that of monophonic wheezes. In the inspiratory phase, pairs of high frequencies perform QPC, whereas in the expiratory phase, the submultiple frequencies of those in the inspiratory phase perform QPC. The harmonics involved result in a polyphonic chord, because apart from the fundamental frequency (f_0), the octave ($2f_0$) and the fifth of the chord ($3f_0$) perform QPC. This result is consistent with the accepted theory that polyphonic wheezes are made up of several dissonant notes starting and ending simultaneously, such as a chord [12].

In the case of random wheezes, the nonlinearity and non-Gaussianity of the process are evident both in inspiratory and expiratory case. In cases of sibilant (SBR) and sonorous (SNR) ronchi, high- and low-frequency pairs perform QPC, because they correspond to high and low wheezes, respectively. From $(f_1, f_2)|_{expiratory\ SNR} = (1/3)(f_1, f_2)|_{inspiratory\ SBR}$, it can be concluded that sibilant and sonorous ronchi have similar nonlinear production mechanism but different transmission, because high frequencies are only amplified in the case of sibilant ronchi.

In the case of stridors, high frequency pairs, both in inspiration and expiration, perform QPC. The frequencies with QPC corresponding to inspiratory stridors from an infant with croup are higher than those corresponding to inspiratory stridors from a 9-year-old child with croup. As well known, stridor intensity is the only thing that distinguishes a stridor from a monophonic wheeze [12]. From the nonlinear analysis, it can be seen that $(f_1, f_2)_{inspiratory\ MW} = (1/5)(f_1, f_2)|_{expiratory\ ST}$ justifying their relationship, but the degree of nonlinearity is much larger in stridor than in monophonic wheeze. This difference can introduce another field of separating the two associated autoregressive musical lung sounds (AMLS).

2.2.6   Autoregressive-HOS Modeling of LS Source and Transmission

Modeling of the system that produces the bioacoustic signals (BAS) contributes to the understanding of their production mechanisms. The manner in which the pathology affects these mechanisms is reflected in the modeling parameters adopted, which is a very important issue in the analysis of BAS, because efficient modeling could lead to an objective description of the changes a disease imposes to the production or transmission path of the BAS.

LS originated inside the airways of the lung are modeled as the input to an all-pole filter, which describes the transmission of LS through the parenchyma and chest wall structure [89]. The output of this filter is considered to be the LS recorded at the chest wall. The recorded LS also contain heart sound interference, the reduction of which is thoroughly addressed in the next chapter. Muscle and skin noise, along with instrumentation noise, are modeled as an additive Gaussian noise. With this model, given a signal sequence of LS at the chest wall, an autoregressive (AR) analysis based on TOS, namely AR-TOS, can be applied to compute the model parameters. Therefore, the source and transmission filter characteristics can be separately estimated, as thoroughly described in [52, 58].

The AR-TOS is described from the following equation:

$$y_n + \sum_{i=1}^{p} a_i y_{n-i} = w_n, \quad a_0 = 1, \tag{2.21}$$

where, y_n represents a pth-order AR process of N samples ($n = 0, \ldots, N - 1$), a_i are the coefficients of the AR model, and w_n are i.i.d., third-order stationary, zero mean, with $E\{w_n^3\} = \beta(0$ and y_n independent of w_l for $n < l$. Because w_n is third-order stationary, y_n is also third-order stationary, assuming it is a stable AR model. For the model of (2.21), we can write the cumulant-based "normal" equations [90]:

$$\sum_{i=0}^{p} a_i R(\tau_1 - i, \tau_2) = 0, \ \tau_1 = 1, \ldots, p \quad \text{and} \quad \tau_2 = -p, \ldots, 0, \tag{2.22}$$

where $R(\tau_1, \tau_2)$ is the third-order cumulant sequence of the AR process. In practice we use sample estimates of the cumulants. Equation (2.22) yields consistent estimates of the AR parameters maintaining the orthogonality of the prediction error sequence to an instrumental process derived from the data [91].

The profound motivations behind the use of the AR-TOS model is the suppression of Gaussian noise, because TOS of Gaussian signals are identically zero. Hence, when the analyzed waveform consists of a non-Gaussian signal in additive symmetric noise (e.g., Gaussian), the parameter estimation of the original signal with TOS takes place in a high-SNR domain, and the whole parametric presentation of the process is more accurate and reliable [86, 92]. The model used for the LS originating inside the airways considers the LS source as the output from an additive combination of three types of noise sequences [17]. The first sequence (periodic impulse) describes the continuous adventitious sound sources, because they have characteristically distinct pitches, and they are produced by periodic oscillations of the air and airway walls [12, 71].

The second sequence (random intermittent impulses) describes the crackle sources, because they are produced by sudden opening/closing of airways or bubbling of air through extraneous liquids in the airways—both phenomena associated with sudden intermittent bursts of sounds energy [12, 71]. Finally, the third sequence (white non-Gaussian noise) describes the breath sound sources, because they are produced by turbulent flow in a large range of airway dimensions [12, 71]. The estimation of the AR-TOS model input (LS source) can be derived from the prediction error via inverse filtering [52, 93].

Two examples of AR-TOS ($p = 2$) modeling (fine crackles from pulmonary fibrosis and squawks from interstitial fibrosis) [58, 94] are shown in Figure 2.2(i) and (ii), respectively. In the case of fine crackles [Figure 2.2(i)], the estimated source waveform [Figure 2.2(i), b] contains impulsive bursts, corresponding to fine crackles [Figure 2.2(i), a]. This could be explained by the associated-with-fine-crackles phenomenon of explosive reopening of small airways that had closed during the previous expiration. The abnormal airway closure that precedes the "crackling" reopening is due to increased lung stiffness [91].

In the case of squawks [Figure 2(ii)], the estimated source [Figure 2.2(ii), b], combines impulsive bursts followed by short, almost exponential, decaying periodic train of impulses. This result is consistent with the accepted theory that squawks are produced by the explosive opening and decaying fluttering of an unstable airway [91]. The transmission filter response in the case of fine crackles [Figure 2.2(i), c] is centered at high frequencies ($\approx$530 Hz), whereas in the case of squawks [Figure 2.2(ii), c], it is centered at lower frequencies ($\approx$380 Hz). From the magnitude of parametric bispectrum of both cases [Figure 2.2(i), d and (ii), d], it can be noticed that pulmonary fibrosis converts the transmission path to a band-pass filter (350–700 Hz), whereas interstitial fibrosis converts it to a rather smoothed low-pass one (cutoff frequency$\approx$480 Hz).

2.3   LOWER-ORDER STATISTICS
2.3.1   Epitomized Rationale
A broad class of non-Gaussian phenomena encountered in practice can be characterized as impulsive. Signals and/or noise in this class tend to produce large-amplitude excursions from the average value more frequently than Gaussian signals. They are more likely to exhibit sharp spikes or occasional bursts of outlying observations than one would expect from normally distributed signals. As a

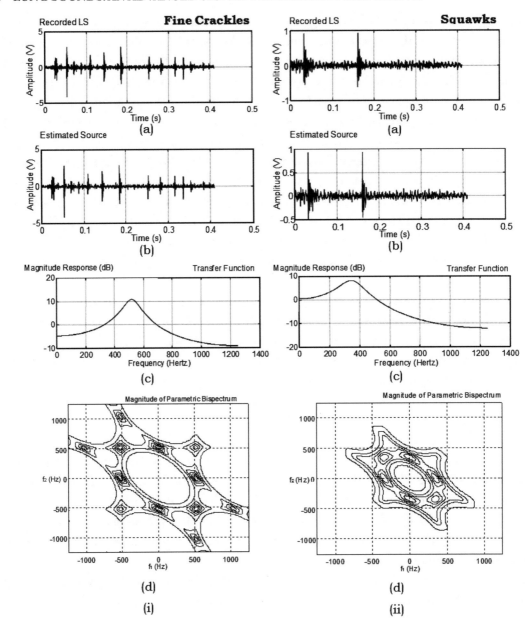

FIGURE 2.2: An example of AR-TOS modeling of LS source and transmission for the case of (i) fine crackles and (ii) squawks [58, 94]. In both columns, (a), (b), (c), and (d) correspond to the recorded LS signal, the AR-TOS-based estimated source, the AR-TOS-based estimated transfer function, and the corresponding parametric bispectrum, respectively. Differences in behavior between the two examined LS in the AR-TOS analysis domain are evident.

result, their density function decay in the tails less rapidly than the Gaussian density function [95]. Among the different types of LS, there are some that seem to belong in the aforementioned class, occurring with a different degree of impulsiveness. Because modeling based on α-stable distribution related to *lower-order statistics* (LOS) is appropriate for enhanced description of impulsive processes [95], its use could provide a useful processing tool for analysis of impulsive LS.

Sustaining the non-Gaussian assumption in the context of analyzing impulsive processes, among the various distribution models that were suggested in the past, the α-stable distribution is the only one that is motivated by the generalized central limit theorem. This theorem states that the limit distribution of the sum of random variables with possibly infinite variances is stable distribution [96]. Stable distributions are defined by the stability property stating that a random variable, X, is stable if and only if the sum of any two independent random variables with the same distribution as X also has the same distribution [96].

Analysis of LS with explosive character, such as crackles, via LOS-based modeling provides an innovating perception in the analysis of impulsive LS, suggesting a new field in their processing for diagnostic feature extraction.

2.3.2   LOS Definitions: α-Stable Distribution

A univariate distribution function $F(x)$ is called α-stable if its characteristic function can be expressed in the following form [95]:

$$\phi(t) = \exp\{ jat - \gamma|t|^\alpha [1 + j\beta \operatorname{sign}(t)\omega(t,\alpha)]\}, \qquad (2.23)$$

where

$$\omega(t,\alpha) = \begin{cases} \tan\frac{\alpha\pi}{2} & \text{for } \alpha \neq 1 \\ \frac{2}{\pi}\log|t| & \text{for } \alpha = 1, \end{cases} \qquad (2.24)$$

$\operatorname{sign}(t) = 1, 0$, and -1 for $t > 0$, $t = 0$, and $t < 0$, respectively, and

$$-\infty < a < \infty, \gamma > 0, 0 < \alpha \leq 2, -1 \leq \beta \leq 1. \qquad (2.25)$$

Thus, the following four parameters can completely determine the α-stable distribution: (1) the location parameter γ, which is the symmetry axis; (2) the scale parameter , also called the *dispersion*, which, in analogy to the variance of the Gaussian distribution, is a measure of the deviation around the mean; (3) the symmetry parameter β, which is the index of skewness; and (4) the characteristic exponent, a, which is a measure of the thickness of the tails of the distribution (a small value of α implies considerable probability mass in the tails of the distribution, whereas a large value of α implies considerable probability mass in the central location of the distribution).

The special cases $\alpha = 2$ and $\alpha = 1$ with $\beta = 0$ correspond to the Gaussian distribution and the Cauchy distribution, respectively. When $\beta = 0$, the distribution is symmetric about the center α, in which case the distribution is called symmetric α-stable (SαS), and its characteristic function is of the form

$$\phi(t) = \exp\left\{jat - \gamma |t|^\alpha\right\} \qquad (2.26)$$

For SαS distributions, α is the mean when $1 < \alpha < 2$ and the median when $0 < \alpha < 1$.

2.3.3   LOS Definitions: Fractional Lower-Order Moments and Covariation Coefficient

Although the second-order moment of an SαS random variable X with $0 < \alpha < 2$ does not exist, all moments of order less than α do exist and are called the *fractional low-order moments* (FLOM), defined as follows:

$$E\left(|X|^p\right) = C(p, \alpha)\gamma^{p/\alpha}, \ 0 < p < \alpha, \ a = 0, \qquad (2.27)$$

where

$$C(p, \alpha) = \frac{2^{p+1}\Gamma\left(\frac{p+1}{2}\right)\Gamma(-p/\alpha)}{\alpha\sqrt{\pi}\,\Gamma(-p/2)}, \qquad (2.28)$$

$E(\cdot)$ denotes the expectation value and $\Gamma(\cdot)$ is the usual gamma function.

The covariation coefficient of jointly SαS random variables X and Y with $\alpha > 1$—which, although it plays the role of correlation coefficient of second-order random variables, can become unbounded—is given by

$$\lambda_{XY} = \frac{E(XY^{\langle p-1 \rangle})}{E(|Y|^p)}, \qquad (2.29)$$

for any $1 \le p < \alpha$ For a thorough description of α-stable distributions, the reader is referred to [95].

2.3.4   LOS Analysis Tools: Converging Variance Test

Because the property that differentiates the Gaussian and non-Gaussian distributions is that non-Gaussian stable distributions do not have finite variance, a test based on the convergence of the variance of the population distribution can lead to the adoption of the right assumption. Specifically, if the population distribution $F(x)$ has finite variance, then the *running sample variance* S_n^2, defined as

$$S_n^2 = \frac{1}{n} \sum_{k=1}^{N} (X_k - \bar{X}_n)^2 \quad 1 \le n \le N,$$

(2.30)

where

$$\bar{X}_n = \frac{1}{n} \sum_{k=1}^{n} X_k,$$

(2.31)

and X_k, $k = 1, \ldots, N$ are samples from the distribution, should converge to a finite value [95].

2.3.5 LOS Analysis Tools: Parameter Estimates for SαS Distributions

Because the three parameters (α, a, and γ) determine the SαS distribution, their estimation from the realizations of a symmetric stable random variable could reflect the differences among various types of impulsive LS.

Although estimation of the parameters of a stable distribution is generally severely hampered by the lack of known closed-form density functions (for all but a few members of the stable family), there are several numerical methods that have been suggested in the literature that can be used to perform reliable parameter estimation: fractile method [97], regression method [98], log|SαS| method [95], and negative-order moment method [95]. A short description of the first three methods is included here. For convenience, γ is replaced by a new parameter, c, defined by

$$c = \gamma^{1/\alpha}.$$

(2.32)

For the symmetric stable law, the fractile estimate

$$\hat{\upsilon}_\alpha = \frac{\hat{x}_{0.95} - \hat{x}_{0.05}}{\hat{x}_{0.75} - \hat{x}_{0.25}},$$

(2.33)

where $\hat{x}_f$ is the f fractile ($0 < f < 1$), which leads to a consistent estimate of α, $\hat{\alpha}$, using tables, such as those in [97], with matched value of $\hat{\upsilon}$. The location parameter a is estimated as the 75% truncated sample mean of the X_k, ($k = 1, \ldots, N$) observations, $\hat{a}_{75}$, that is, as the arithmetic mean of the middle 75% of the ranked observations. A consistent estimator of c can be found by the following equation:

$$\hat{c} = \frac{\hat{x}_{0.75} - \hat{x}_{0.25}}{\upsilon_c(\hat{\alpha})},$$

(2.34)

using tables, such as those in [97], for calculation of $\hat{\upsilon}_c(\hat{\alpha})$.

The regression method [98] uses the fractile parameter estimates as initial estimates and proceeds iteratively until the change in the estimated values between two successive iterations, *acc*, is sufficient small. The new estimates can be found by solving (via least-squares) the following over-determined system of linear equations:

$$y_l = \mu + \alpha w_l \quad l = 1, \dots, L, \tag{2.35}$$

where

$$w_l = \log|t|, \; t = (\pi l/25), \tag{2.36}$$

$$y_l = \log\left(-2\log\left|\frac{1}{N}\sum_{k=1}^{N}\exp\left(jtX'_k\right)\right|\right), \tag{2.37}$$

$$X'_k = (X_k - \hat{a}_{75})/\hat{c}_i, \quad k = 1, \dots, N, \tag{2.38}$$

$$\mu = \log(2c^\alpha). \tag{2.39}$$

The log|SαS| method [95] uses the mean, $E(Y)$, and the variance, $\mathrm{Var}(Y)$, of a new random variable $Y = \log|X|$, for the estimation of α and γ as follows:

$$\hat{\alpha} = \left(1\Big/\left(\frac{6\mathrm{Var}(Y)}{\pi^2} - 0.5\right)\right)^{1/2} \hat{\gamma} = \exp\left(\hat{\alpha}\left(E(Y) - C_e\left(\frac{1}{\hat{\alpha}} - 1\right)\right)\right), \tag{2.40}$$

where $C_e = 0.57721566\dots$ is the Euler's constant.

2.3.6   LOS Analysis Tools: AR Modeling of SαS Distributions

A pth-order AR SαS process is described by the following equation:

$$X(n) = \sum_{i=1}^{P} a_i X(n-i) + U(n), \tag{2.41}$$

where $\{U(n)\}$ is a sequence of i.i.d. SαS random variables of characteristic exponent α and dispersion Y_u. Furthermore, the stationary process $X(n)$ is SαS and $X(n)$ and $U(n+j)$ are independent for any $j > 0$. From the several proposed methods of estimating α_i ($i = 1, \dots, P$) from observations of the output $X(n)$, the least-squares (LS) method [99] and the method that combines FLOM with the Yule–Walker (YW) equation [95] are usually adopted.

In the LS case, the AR model is treated as if it is driven by a second-order process and thus the linear least-squares (LSq) regression can be used. In that case, the LS estimates of α_i $(i \, \hat{a} \, 1, ..., P)$ are the solution of the following equation:

$$\hat{\mathbf{C}}_{\text{LS}}\hat{\mathbf{a}}_{\text{LS}} = \hat{\mathbf{b}}_{\text{LS}} \qquad (2.42)$$

where $\hat{\mathbf{C}}_{\text{LS}} = (\hat{\lambda}(i,j))$, $\hat{\mathbf{b}}_{\text{LS}} = (\hat{\lambda}(i))$, and for sufficient large data length, $\hat{\lambda}(i,j) \approx \hat{\lambda}_{\text{LSq}}(1-j)$ and $\hat{\lambda}(i) \approx \hat{\lambda}_{\text{LSq}}(i)$, where $\hat{\lambda}_{\text{LS}}(i)$ is the LSq estimate of the covariation of $X(n+i)$ and $X(n)$.

In the case of YW equation using FLOM, the coefficients of the AR system can be found by solving the following system of YW linear equations:

$$\mathbf{Ca} = \mathbf{b}, \qquad (2.43)$$

where $b = [\lambda(1)\lambda(2)...\lambda(P)]^{\text{T}}$, $a = [a_1 \, a_2 \, ... \, a_P]^{\text{T}}$ and

$$\mathbf{C} = \begin{bmatrix} \lambda(0) & \lambda(-1) & \cdots & \lambda(1-P) \\ \lambda(1) & \lambda(0) & \cdots & \lambda(2-P) \\ \vdots & \vdots & \ddots & \\ \lambda(P-1) & \lambda(P-2) & \cdots & \lambda(0) \end{bmatrix}, \qquad (2.44)$$

where

$$\lambda(k) = E\left\{X(n)X(n-k)^{\langle p-1 \rangle}\right\} \Big/ E\left\{|X(n-k)|^p\right\}, \qquad (2.45)$$

and $p \geq 1$ is arbitrary.

2.3.7   LOS Analysis of Discontinuous Adventitious Sounds

Following this approach, impulsive LS [i.e., discontinuous adventitious sounds (DAS)] with explosive character can be modeled via LOS [76]. From the estimated parameters of the SαS distribution of the analyzed impulsive LS using the log|SαS| method [95], it is derived that the covariation coefficient $\lambda_{\text{SQ-FC}}$ calculated for the cases of sound sources of squawks (SQ) and fine crackles (FC) [76] shows an almost 50% correlation between the SQ and FC, confirming the accepted theory that SQ are produced by the explosive opening, due to an FC, and decaying fluttering of an unstable airway [12, 71].

Figure 2.3 depicts the estimated LS source for the case of FC, coarse crackles (CC), and SQ when using the LOS-based AR modeling of (2.40). In fact, by using inverse filtering, an estimation of the input $U(n)$ can be found (Figure 2.3a–c for FC, CC, and SQ, respectively) [76]. Visual inspection of the findings confirms the explosive character of the source sound of impulsive LS, because they are produced by sudden changes in the acoustic properties of the lung.

By estimating the characteristic exponent α for each category (crackles, background noise, and artifacts), using the log|SαS| method [95], a classification criterion could be established based on the estimated α values. This approach has been adopted by Hadjileontiadis et al. [100], who deduced that FC follow approximately a Cauchy distribution (mean $\alpha = 0.92$, which is close to $\alpha = 1.0$ that holds in the case of Cauchy distribution), whereas the CC deviate both from Cauchy and Gaussian distributions (mean $\alpha = 1.33$; the artifacts have high impulsiveness with values of $\alpha \ll 1.0$). These results are illustrated in Figure 2.4. The events that resemble the vesicular sound are seen as background noise and are modeled as Gaussian processes (values of $\alpha > 1.8$ or very close to 2.0, which holds in the case of Gaussian distribution). Using this approach, disputed sounds that could not be classified by the physician (marked as A?) are clearly classified by the proposed method, as shown in Figure 2.4e.

The above examples show that SαS distribution- and LOS-based modeling of impulsive LS provides a measure of DAS impulsiveness, and, at the same time, reveals the underlying relationships between the associated production mechanisms and pathology.

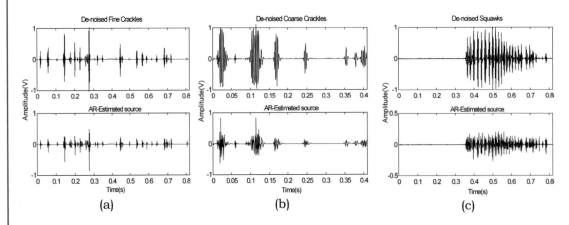

FIGURE 2.3: Estimation of the LS source sound for recorded DAS. In all subfigures, the upper and lower parts denote the recorded DAS (after denoising) and their estimated source using AR-LOS modeling. (a) Fine crackles, (b) coarse crackles, and (c) squawks [76].

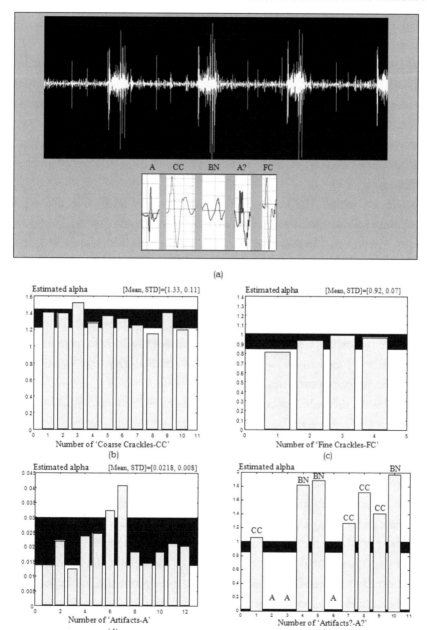

FIGURE 2.4: (a) Analyzed signal [100]. (b–d) Estimated α for crackles and artifacts. (e) Classification results of disputed peaks (A?) using SαS modeling. FC, fine crackles; CC, coarse crackles; A, artifacts; BN, background noise.

2.4   WAVELET ANALYSIS

2.4.1   Epitomized Rationale

The notion of a wavelet (i.e., a small wave) has evolved rapidly from its introduction by Grossmann and Morlet [101] in the mid 1980s as applied to the analysis of properties of seismic and acoustic signals. Nowadays, the family of analyzing functions dubbed wavelets is being increasingly used in problems of pattern recognition; in processing and synthesizing various signals (e.g., speech); in analysis of images of any type (e.g., iris images, X-rays, satellite images, an image of mineral, etc.); for study of turbulent fields, for contraction (compression) of large volumes of information, and in numerous other cases.

The wavelet transform (WT) of a one-dimensional signal involves its decomposition over a basis obtained from a soliton-like function (wavelet), possessing some specific properties, by dilations and translations. Each of the functions of this basis emphasizes both a specific spatial (temporal) frequency and its localization in physical space (time). Thus, unlike the Fourier transform tradition-ally used in signal analysis, WT offers a two-dimensional expansion of a given one-dimensional signal, with the frequency and the coordinate treated as independent variables. Consequently, the signal could be simultaneously analyzed in physical (time, coordinate) and frequency spaces. This ability has spawned a number of sophisticated wavelet-based methods for signal manipulation and interrogation. WT was found to be particularly useful for analyzing signals that can best be de-scribed as aperiodic, noisy, intermittent, and transient [102].

One faces known difficulties when processing short high-frequency signals, or signals with localized frequencies. WT proves to be an extremely efficient tool in adequately decoding such sig-nals because elements of its basis are well localized and possess a moving time–frequency window. It is not a coincidence that many researchers refer to wavelet analysis as a "mathematical microscope," because this term accurately conveys the remarkable capability of the method to offer a good resolu-tion at different scales. This so-called microscope reveals the internal structure of an essentially in-homogeneous process (or field) and exposes its local scaling behavior [103]. In this vein, WT picks out "coherent structures" in a time signal at various scales by shifting the wavelet along the signal, and so coherent structures related to a specific dilation in the signal to be identified.

Wavelets can be successfully applied to solve various problems, as the vast relevant litera-ture justifies. For a more detailed description about WT, the reader should refer to the works of Daubechies [104], Farge [105], and Mallat [106].

2.4.2   Continuous Wavelet Transform

WT analysis uses wavelets to convolve them with the signal under investigation [102]. The WT of a continuous signal with respect to the wavelet function is defined as [102]

$$W_s(a,b) = \int_{-\infty}^{\infty} s(t)\psi_{a,b}^*(t)dt, \tag{2.46}$$

where $s(t)$ is the time-domain signal ($s(t) \in L^2(\mathfrak{R})$), $*$ is the complex conjugate, $\psi_{a,b}(t)$ is the mother wavelet scaled by a factor a, ($a > 0$), and dilated by a factor b, that is,

$$\psi_{a,b}(t) = \frac{1}{\sqrt{a}}\psi\left(\frac{t-b}{a}\right). \tag{2.47}$$

Substituting (2.47) in (2.46), we get the definition of the continuous wavelet transform (CWT) as

$$W_s(a,b) = \frac{1}{\sqrt{a}}\int_{-\infty}^{\infty} s(t)\psi^*\left(\frac{t-b}{a}\right)dt. \tag{2.48}$$

For $b = 0$, (2.48) takes the form of

$$W_s(a) = \frac{1}{\sqrt{a}}\int_{-\infty}^{\infty} s(t)\psi^*\left(\frac{t}{a}\right)dt. \tag{2.49}$$

If we assume $s(t)$ to be a sine wave (e.g., a musical LS signal, i.e., wheeze), then

$$s(t) = A\sin(\omega_s t + \varphi), \tag{2.50}$$

where ω_s and ϕ are the angular frequency and phase of the sine wave, respectively, and set $\omega_s = \lambda\omega_c/a$, where ω_c/a is the angular frequency of the analyzing mother wavelet with ω_c being its central angular frequency ($a = 1$), and $\lambda = \omega_s/(\omega_c/a)$, then, assuming $\psi \in \mathfrak{R}$, (2.50) becomes:

$$W_s(a) = \frac{A}{\sqrt{a}}\int_{-\infty}^{\infty} \sin(\lambda\omega_c t/a + \varphi)\,\psi\left(\frac{t}{a}\right)dt, \tag{2.51}$$

which, after using an intermediate parameter of $k = t/a$, it can be simplified in the form

$$W_s(a) = A\sqrt{a}\int_{-\infty}^{\infty} \sin(\lambda\omega_c t + \varphi)\psi(t)dt. \tag{2.52}$$

Inverse CWT is defined as [102]

$$x(t) = \frac{1}{C_g} \int\limits_{-\infty}^{\infty} \int\limits_{0}^{\infty} W_s(a,b) \psi_{a,b}(t) \frac{dadb}{a^2},$$

(2.53)

where C_g is an admissibility constant [102], and its value depends on the chosen wavelet.

2.4.3 Discrete WT and Multiresolution Representation

By selecting discrete values for a, b parameters, a discrete form of the wavelet of (2.47) is produced. When a power-of-two logarithmic scaling of both the dilation and translation steps is adopted (known as the "dyadic grid arrangement"), a dyadic grid wavelet is formed,

$$\psi_{m,n}(t) = 2^{-m/2} \psi \left(2^{-m}t - n\right),$$

(2.54)

where m and n are integers that control the wavelet dilation and translation, respectively. Discrete dyadic grid wavelets are commonly chosen to be orthonomal; they are both orthogonal to each other and normalized to have unit energy. Using the dyadic grid wavelet of (2.54), discrete WT (DWT) can be defined as

$$T_{m,n} = \int\limits_{-\infty}^{\infty} x(t) \psi_{m,n}(t) dt,$$

(2.55)

along with the inverse DWT,

$$x(t) = \sum\limits_{m=-\infty}^{\infty} \sum\limits_{n=-\infty}^{\infty} T_{m,n} \psi_{m,n}(t),$$

(2.56)

requiring the summation over all integer m and n.

Orthonormal dyadic discrete wavelets are associated with scaling function and their dilation equations [102]. The scaling function is associated with the smoothing of the signal and is given by

$$\varphi_{m,n}(t) = 2^{-m/2} \varphi(2^{-m}t - n).$$

(2.57)

The scaling function is orthogonal to translations of itself, yet not to dilations of itself. The convolution of the scaling function with the signal produces approximation coefficients, that is,

$$S_{m,n} = \int\limits_{-\infty}^{\infty} x(t) \varphi_{m,n}(t) dt,$$

(2.58)

that are simply weighted averages of the continuous signal factored by $2^{\frac{m}{2}}$.

The signal $x(t)$ can be represented using a combined series expansion using both the approximation coefficients and the wavelet (detail) coefficients in the following form:

$$x(t) = \sum_{n=-\infty}^{\infty} S_{m_0,n}\varphi_{m_0,n}(t) + \sum_{m=-\infty}^{m_0} \sum_{n=-\infty}^{\infty} T_{m,n}\psi_{m,n}(t),$$

(2.59)

revealing that the original signal can be expressed as a combination of an approximation of itself, at arbitrary scale index m_0, added to a succession of signal details from scales m_0 down to negative infinity. By denoting the signal detail at scale m as

$$d_m(t) = \sum_{n=-\infty}^{\infty} T_{m,n}\psi_{m,n}(t).$$

(2.60)

Equation (2.59) can be written in the form of

$$x(t) = x_{m_0}(t) + \sum_{m=-\infty}^{m_0} d_m(t),$$

(2.61)

hence,

$$x_{m-1}(t) = x_m(t) + d_m(t).$$

(2.62)

Equation (2.62) represents the *multiresolution representation*, because it tells us that if we add the signal detail at an arbitrary scale (index m) to the approximation at that scale we get the signal approximation at an increased resolution, that is, at a smaller scale (index $m - 1$) [102].

The above representation can be realized through the concept of vector spaces. For example, in the case of an image analysis, for each vector space, there is another vector space of higher resolution until you get to the final image. Also, each vector space contains all vector spaces that are of lower resolution. The basis of each of these vector spaces is the scale function for the wavelet. For practical purposes, one can think of an image as a vector space—such that V^j would be the perfectly normal image and V^{j-1} would be that image at a lower resolution until you get to V^0, where you just have 1 pixel in the entire image. For each vector space V^j, there is an orthogonal complement called W^j and the basis function for this vector space is the wavelet.

The decomposition of the input signal into approximation and detail space is called *multiresolution approximation* [107], which can be realized using a pair of finite impulse response filters (and their adjoints), which are low pass and high pass, respectively, defining a *multiresolution decomposition–multiresolution reconstruction* scheme (MRD–MRR).

An analytical description and details regarding the implementation of wavelet analysis can be found in [106–111] and in the references contained therein.

2.4.4   Wavelet-based Analysis of LS

Considering the characteristics of various LS signals, a series of research efforts were developed toward the application of WT to LS analysis. In the work of Gross et al. [112], parameters based on multiresolution approximation were able to detect typical pneumonia LS at an early stage of the disease in all 16 examined patients. In particular, using Daubechies-8 coefficient quadratic mirror filters [108], a multiresolution approximation was performed and the bronchial breathing sound pattern was detected by using WT-based ratio R between inspiration and expiration of the frequency band 345–690 Hz (scale 3) described in [113].

Figure 2.5 illustrates the distribution of the WT coefficients at scale 3 during the breathing cycle (Figure 2.5a) for the healthy (Figure 2.5b) and the pneumonia (Figure 2.5c) sites for a patient with bronchial breathing and no crackles. As shown in this figure, WT coefficients in the pneumonia site exhibit reduced amplitude compared to the ones from the healthy site, following a reversed pattern (healthy site: higher amplitude in inspiration and lower in expiration; pneumonia site: lower amplitude in inspiration and higher in expiration).

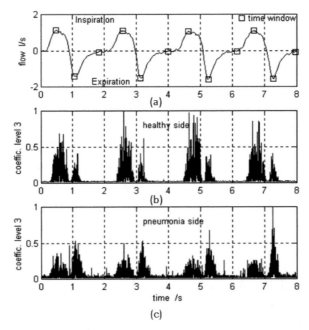

FIGURE 2.5: WT-based analysis of LS from a patient with pneumonia. (a) Breathing pattern, (b) WT coefficients from scale 3 of LS recorded from the healthy site, and (c) WT coefficients from scale 3 of LS recorded from the pneumonia site [112].

The successful WT-based detection of bronchial breathing can be considered as a first step in developing a monitoring system for patients at risk for pneumonia.

Ayari et al. [114] applied the WT transform to normal LS, crackles, and wheezes. In their work, they used wavelets formed by the first derivative of a Gaussian with standard deviation of $\sigma = 32 \times 10^{-5}$ s. Derivatives of Gaussian are most often used to guarantee that all maxima lines propagate up to the finest scale. They used the WT modulus maxima method [115] to characterize the local regularity of LS. The localized singularities correspond to the sharp and slow variations. WT of LS across scales presents local maxima when using a wavelet that is a first derivative of a smoothing function. Consequently, the sharp and slow variations are indicated by the location of the local modulus maxima obtained by one vanishing moment wavelet. Analysis of the behavior of WT modulus maxima across scales permits detection of singularities and estimation of Lipschitz exponents [116] for the sharp and slow variation. Numerical results for the crackle sound showed that singularity at sharp variation was Lipschitz −1. For wheezes, the singularity at sharp variations was Lipschitz 1. This is quite expected, and as wheezing sound seems more like a sinusoidal signal, it appears to be more regular than the normal LS.

WT has also been used in combination with artificial neural networks (ANN) to form an LS classification scheme. In the work of Kandaswamy et al. [117], LS signals were decomposed into the frequency subbands using multiresolution approximation, and a set of statistical features was extracted from the subbands to represent the distribution of wavelet coefficients. An ANN-based system, trained using the resilient back-propagation algorithm (RPROP) [118], was implemented to classify LS into six categories: normal, wheeze, crackle, squawk, stridor, or rhonchus. RPROP is a local adaptive learning scheme, performing supervised batch learning in feed-forward neural networks. The basic principle of RPROP is to eliminate the harmful influence of the size of the partial derivative on the weight step. As a consequence, only the sign of the derivative is considered to indicate the direction of the weight update.

The authors used some statistics from the WT coefficients as features for the ANN-based classification. In particular, they used: (1) the mean of the absolute values of the coefficients in each subband, (2) the average power of the wavelet coefficients in each subband, (3) the standard deviation of the coefficients in each subband, and (4) the ratio of the absolute mean values of adjacent subbands. The first two features represent the frequency distribution of the signal, whereas the next two denote the amount of changes in frequency distribution. These feature vectors, calculated for the frequency scales of 3–7, were used for classification of LS signals by ANN. The authors achieved classification accuracy of >90% when using Daubechies wavelet of order 8 [108].

As is apparent from the examples cited above, WT provides new opportunities in LS representation that allow construction of hybrid analysis tools. The latter stem from the beneficial advantages of WT and, by combining other efficient methods, introduce new approaches in LS analysis and appreciation. This will become clearer in the succeeding section.

2.5   WAVELET-HOS

2.5.1   Epitomized Rationale

Wavelet analysis can be seen as a generalization of the Fourier analysis and in many cases permits a similar interpretation, but amplifies it by adding time resolution—in a more fundamental way than is permitted by the short-time Fourier transform (STFT), because the latter does not remove the objection raised above against Fourier-type methods.

Signals containing coherent couplings have traditionally been analyzed by means of a normalized bispectrum, that is, the bicoherence [see (2.12) and (2.13)]. The bicoherence, when defined properly, quantifies the fraction of power contained in the nonlinearity. However, when the available data have a nonstationary nature, traditional fast Fourier transform (FFT)-based methods [91, 119] may be inadequate due to the inability of the STFT to resolve short-lived transients properly. To circumvent such problem, the wavelet bispectrum (WBS)/bicoherence is proposed [120–122], based on a combination of CWT with HOS.

For the analysis of nonstationary processes, the WBS has two main advantages compared to traditional FFT-based methods. First, because CWT is a timescale representation of a signal, thus, a time axis is introduced in a natural way. Second, wavelets have an inherent constant-Q filtering property, and are consequently well suited for detection of transients.

The wavelet bicoherence (WBC) technique detects QPC while reducing time averages to a minimum, thus permitting short-lived events, pulsed, and intermittency to be resolved [123]. Relatively short data sequences are sufficient to perform an analysis, in contrast to the Fourier bicoherence, which needs long time series to obtain both sufficient frequency resolution and statistics. Estimates of the noise contribution and error level of the WBC provide a criterion for the reliability of the results. A powerful noise reduction is an integral part of the standard technique: as noncoherent contributions are averaged out, weak coherent signals can be detected in very noisy data. Moreover, the wavelet bicoherence is more independent of the frame of reference and may be expected to be more useful in experiments where measurement in the local frame of reference is difficult [120].

2.5.2   CWT-HOS Definitions

Wavelet bispectrum. By analogy to the definition of the bispectrum in Fourier terms [see (2.10)], WBS is defined as [123]

$$B_w(a_1, a_2) = \int_T W_x^*(a, \tau) W_x(a_1, \tau) W_x(a_2, \tau) \, d\tau,$$

(2.63)

where $W_x(a_1, \tau)$, $i = 1, 2$ denotes that the CWT defined in (2.46) and the integration is performed over a finite time interval T: $\tau_0 \le \tau \le \tau_1$, and a, a_1, a_2 satisfy the following rule:

$$\frac{1}{a} = \frac{1}{a_1} + \frac{1}{a_2}.$$

(2.64)

The WBS expresses the amount of QPC in interval T, which occurs between wavelet components of scale lengths a_1, a_2, and a of $x(t)$ such that the sum rule of (2.64) is satisfied. By interpreting the scales as inverse frequencies, $\omega = 2\pi/a$, the WBS can be interpreted as the coupling between wavelets of frequencies that satisfy $\omega = \omega_1 + \omega_2$, within the frequency resolution.

Wavelet bicoherence. Similarly to the definition of bicoherence [see (2.12)], WBC can be defined as the normalized WBS, that is,

$$b_w(a_1, a_2) = \frac{B_w(a_1, a_2)}{\left\{ \left[\int_T |W_x(a_1, \tau) W_x(a_2, \tau)|^2 \, d\tau \right] \left[\int_T |W_x(a, \tau)|^2 \, d\tau \right] \right\}^{1/2}},$$

(2.65)

in which magnitude $|b_w(a_1, a_2)|$, namely, wavelet bicoherence index, can attain values between 0 and 1. For ease of interpretation, the *squared WBC* plotted in the (ω_1, ω_2) plane, that is, $|b_w(\omega_1, \omega_2)|^2$, is preferred. Because of the symmetries in the definition and the limitation set by the Nyquist frequency ω_s [124], the estimation of WBC in the whole bifrequency plane can be based on its values in the principal region $\{\Delta: \omega_1, \omega_2 \le \omega_1, \omega_1 + \omega_2 \le \omega_s\}$.

Summed wavelet bicoherence. For comparing cases computed under the same numerical conditions, the summed wavelet bicoherence (SWBC) could be introduced as

$$b_w^2(\omega) = \sum_\Delta b_w^2(\omega_1, \omega_2).$$

(2.66)

In general, the numerical values of SWBC depend on the chosen calculation grid; thus, they basically provide qualitative summarization of the underlying information.

As van Milligen et al. [123] note, the use of nonorthogonal wavelets, such as Morlet, results in nonstatistically independent wavelet coefficients. This introduces a statistical noise level in the estimation of WBC, with an upper bound given by [123]

$$N_b\left(\omega_1,\omega_2\right) \approx \left[\frac{\pi}{\min\left(|\omega_1|,|\omega_2|,|\omega_1+\omega_2|\right)}\frac{1}{T}\right]^{1/2}.$$

$$(2.67)$$

From (2.67), it can be easily observed that the statistical noise affects the low frequencies of WBC, whereas at higher frequencies it drops rapidly with T. This reveals the ability of WBC to serve as a noise filter for coherent signals.

Evolutionary wavelet bicoherence. Because the WBC defined in (2.65) refers to a certain time interval T, its value is made to correspond to the center of this interval, that is, $t_0 = T/2$. Consequently, evolutionary WBC (EWBC) can be defined as

$$\mathbf{b}_w\left(\omega_1,\omega_2,t\right) = \left\{b_w\left(\omega_1,\omega_2\right)|_{t=t_0+k\Delta T_1}\right\}, \ k = 0,1,2,3,\dots;$$
$$\left(2\pi/\omega_s\right) \le \Delta T_1 \wedge k\Delta T_1 \le T_{\text{total}} - 2t_0,$$

$$(2.68)$$

where T_{total} is the total time duration of the analyzed signal $x(t)$.

When using EWBC, the evolution of the nonlinearities across time can be represented, within a time-resolution controlled by the selection of the ΔT_1 value.

2.5.3 Wheeze Analysis With CWT-HOS

Because of the nature of wheezes, the notion of QPC detection examined in Section 2.2 is expanded by using wavelet bispectrum and wavelet bicoherence as a means to track and quantify the evolution of the nonlinear characteristics of wheezes within the breathing cycle. To this end, the combination of WT with third-order statistics/spectra introduces the nonlinear analysis of wheezes in the time–bifrequency domain. This was investigated by Taplidou and Hadjileontiadis [125], who analyzed breath sound signals from asthmatic patients by using CWT-HOS-based parameters.

Figure 2.6a depicts one breathing cycle of a breath sound signal along with the normalized airflow (superimposed with a dotted line) recorded from an asthmatic patient. The positive and negative airflow values indicate inspiratory and expiratory phases, respectively. As Figure 2.6a shows, the expiratory phase is prolonged compared to the inspiratory one; this is due to the existence of asthma. Similarly, the breath sound signal exhibits a profound high amplitude section (~1.2–2.8 s) corresponding to inspiratory wheeze, whereas an extended expiratory wheeze (~3.7–6.3 s) with decaying amplitude dominates the expiratory phase.

Figure 2.6b shows the representation of the signal in the time–frequency domain via the CWT of the signal. In this figure, the harmonic character of wheezes (both inspiratory and expiratory) is apparent. It is clear that there are coexisting distinct spectral peaks within the area of 150–500 Hz that emerge during the appearance of wheezes, revealing their polyphonic character (such as a chord). In addition, a frequency sweep from low to high frequencies and vice versa can

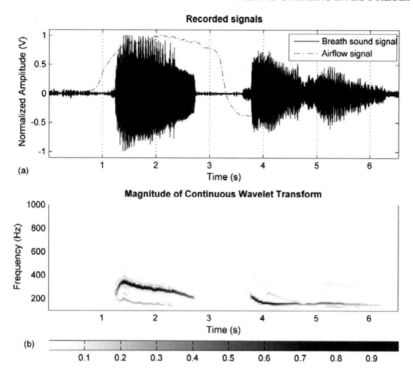

FIGURE 2.6: (a) One breathing cycle of a breath sound recording from an asthmatic patient with two dominant wheezes (one inspiratory and one expiratory). (b) CWT of the analyzed signal [125].

be noticed, mainly at the beginning and the end of wheezes, because of the increase or decrease in airflow signal, respectively. Moreover, the spectral peak within 200–400 Hz that sustains its high amplitude during the inspiratory wheeze subsides in the expiratory one, where the spectral peak about 200 Hz seems to be the most evident one.

The variation in time–frequency characteristics of wheezes justifies the necessity of using wavelet-based HOS to reveal their nonlinearities.

The ability of WBS and WBC to capture the existence of nonlinearities in wheezes is shown in Figure 2.7. In particular, the estimated magnitude of WBS (top), WBC (middle), and squared WBC (bottom) are illustrated in Figure 2.7(i) and (ii), for a time section without wheeze (2.75–3.75 s) and with wheeze (3.75–4.75 s), respectively, corresponding to sections from the breath sound signal depicted in Figure 2.6a. From the comparison of the estimated WBS in Figure 2.7(i) and (ii), it is apparent that the WBS of the section without wheeze [Figure 2.7(i), top] is spread in the area of low frequencies (100–150 Hz) and does not exhibit any distinct peak at a higher-frequency range; furthermore, its values are significantly lower compared to those of the section with wheeze

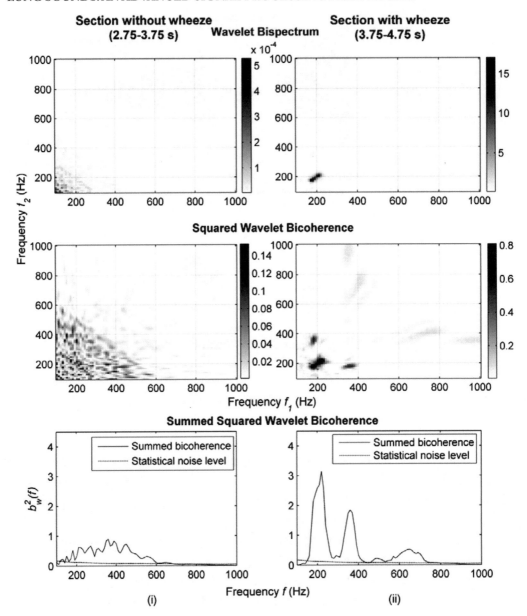

FIGURE 2.7: (i) Section without wheeze (2.75–3.75 s). (ii) Section with wheeze (3.75–3.75 s), with reference to the recorded LS signal depicted in Figure 2.6a. In both cases, WBS (top), squared WBC (middle), and SWBC along with statistical noise level (⋯⋯) (bottom) are depicted accordingly [125].

[Figure 2.7(ii), top]. In the latter, the WBS reveals a concentration of its values around a peak located approximately at $(f_1, f_2) \approx (200, 200)$ Hz. This implies that a possible quadrature self-phase-coupling exists at $(f_1, f_2) \approx (200, 200)$ Hz related to the frequency located at $f_3 = f_1 + f_2 \approx$ 400 Hz. Clearly, the inspection of CWT in Figure 2.6b at the selected section justifies the existence of f_3.

Looking at the estimated magnitude of the squared WBC for both cases [Figure 2.7(i) and (ii), middle], a more enhanced resolution of the bifrequency content is provided. In Figure 2.7(i) middle, a wider spread is noticed, spanning from 100 to 600 Hz, whereas in Figure 2.7(ii) middle, additional distinct peaks appear at higher frequencies. Comparing the values of squared WBC in Figure 2.7(i) and (ii) middle, it can be seen that they are much lower in the time section that is without wheeze (maximum value of 0.15) than those of the time section that contains wheeze (maximum value of 0.8). This shows that practically there is no QPC in the WBC content corresponding to the breath sound signal that does not contain wheeze; in contrast, a strong QPC exists between the harmonics revealed in the WBC content corresponding to the breath sound signal that contains wheeze.

The SWBC, estimated for both cases, is shown in Figure 2.7(i) and (ii) bottom, respectively. The SWBC is plotted along with the upper bound of the statistical noise level (dotted line) defined in (2.67), which in all cases is significantly lower than the estimated SWBC. From the two sub-figures, it is apparent that the SWBC of the breath sound without wheeze [Figure 2.7(i), bottom] sustains a low value at a frequency range of 100–600 Hz, whereas the SWBC of the breath sound with wheeze [Figure 2.7(ii), bottom] demonstrates high peaks of about 200 and 350 Hz and a lower one of about 650 Hz, showing a clear difference from the previous case.

Figure 2.8 illustrates a graphical representation of the ESWBC for whole breath sound signal shown in Figure 2.6a. In this way, the evolution of the nonlinearities in breath sound signal that contains wheezes is demonstrated. From this figure, it is apparent that QPC occurs in both breathing phases only at the time instances where the wheezes exist. Moreover, the main frequency pair with QPC $[(f_1, f_2) \approx (200, 200)$ Hz] is sustained both in inspiratory and expiratory wheezes; however, additional pairs at higher frequencies with QPC emerge during the expiratory wheeze. This relates to the pathology of asthma, because it affects the expiratory phase more than the inspiratory one of asthmatic patients [126].

The CWT-HOS-based analysis of wheezes presented here could be expanded to different types of wheezes as a means to characterize them according to the nonlinear properties they exhibit as they evolve within the breathing cycle. This would shed light on the differences between different categories of wheezes (e.g., monophonic vs. polyphonic) and different pathologies (e.g., COPD and asthma). In either case, CWT-HOS provides an efficient hybrid analysis tool that allows enhance-

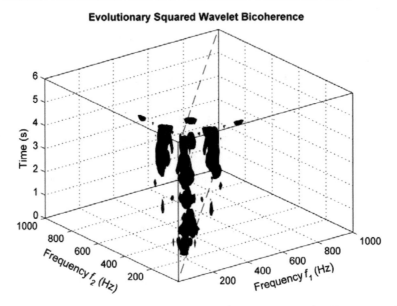

FIGURE 2.8: ESWBC with reference to the recorded LS signal depicted in Figure 2.6a. The time axis includes one breathing cycle [125].

ment of significant details in the LS interpretation, which could not be perceived by auscultation solely.

2.6 HIGHER-ORDER CROSSINGS

2.6.1 Epitomized Rationale

The LS analysis domains that we have presented so far refer to transforms of the initial form of the signal, that is, its time-domain representation, to a new one (e.g., HOS, LOS, CWT domains). In this section, time characteristics of LS would be considered as the main focus. Taking into consideration that almost all observed time series are oscillatory, displaying local and global up and down movements as time progresses, proper time series analysis could "capture" these oscillatory patterns and use them as a new domain of representation. This is done here by using the concept of higher-order crossings (HOC) [127]. HOC analysis of time series is a way to study oscillation of stochastic processes combinatorially by counting and provides an alternative to commonly used spectral methods.

Time ordering makes it possible to apply to a time series linear filtering, one of the most indigenous elements of time series analysis. This filtering, in fact, changes the time series oscillations and this can be reflected in the change in the zero-crossing count effected by the filtering procedure. The application of a specific sequence of filters (or family of filters) to a time series forms the corresponding sequence (or family) of zero-crossing counts (namely, HOC) and, thus, provides a summary of the oscillation "history" observed in the time series and its filtered versions.

Following the properties of HOC estimated from different LS signals, new discrimination features could be established that provide successful differentiation among LS with similar acoustic behavior.

2.6.2   HOC Definitions

For a finite, wide-sense stationary, zero-mean series $\{Z_t\}$, $t = 1, \ldots, N$, its oscillatory pattern about level zero can be expressed through the zero-crossing count. By assuming the iterative procedure of applying a filter to the time series, and counting the number of zero crossings in the filtered time series, applying yet another filter to the original time series, and again observing the resulting zero crossings and so on (filtering and counting), the resulting zero-crossing counts, that is, HOC, could be produced [127]. When a specific sequence of filters is applied to a time series, the corresponding sequence of zero-crossing counts is obtained, resulting in the so-called HOC sequence. Many different types of HOC sequences can be constructed by appropriate filter design, according to the desired spectral and discrimination analysis (DA).

Let ∇ be the backward difference operator defined by

$$\nabla Z_t \equiv Z_t - Z_{t-1}. \tag{2.69}$$

The difference operator ∇ is a high-pass filter. If we define the following sequence of high-pass filters

$$\Im_k \equiv \nabla^{k-1}, \ k = 1, 2, 3, \ldots, \tag{2.70}$$

with $\Im_1 \equiv \nabla^0$ being the identity filter, and with a transfer function given by

$$H(\omega; k) = (1 - \exp(-j\omega))^{k-1}, \tag{2.71}$$

we can estimate the corresponding HOC, namely, simple HOC [127], by

$$D_k = \text{NZC}\{\Im_k(Z_t)\}, \ k = 1, 2, 3, \ldots; \ t = 1, \ldots, N, \tag{2.72}$$

where NZC$\{\cdot\}$ denotes the estimation of the number of zero crossings and

$$\nabla^{k-1} Z_t = \sum_{j=1}^{k} \binom{k-1}{j-1} (-1)^{j-1} Z_{t-j+1} \ \text{with} \ \binom{k-1}{j-1} = \frac{(k-1)!}{(j-1)!(k-j)!}. \tag{2.73}$$

In practice, we only have finite time series and lose an observation with each difference. Hence, to avoid this effect we must index the data by moving to the right, that is, for the evaluation of k simple HOC, the index $t = 1$ should be given to the kth or a later observation. For the estimation of the number of zero crossings in (2.72), a binary time series $X_t(k)$ is initially constructed given by

$$X_t(k) = \begin{cases} 1 \text{ if } \Im_k(Z_t) \geq 0 \\ 0 \text{ if } \Im_k(Z_t) < 0 \end{cases}, \quad k = 1, 2, 3, \ldots; \quad t = 1, \ldots, N, \tag{2.74}$$

and the desired simple HOC are then estimated by counting symbol changes in $X_1(k), \ldots, X_N(k)$, that is,

$$D_k = \sum_{t=2}^{N} [X_t(k) - X_{t-1}(k)]^2. \tag{2.75}$$

From (6), it follows that if $X_{t-1}(k) \neq X_t(k)$, then $X_t(k+1) = X_t(k)$. Thus, every symbol change in $\{X_t(k)\}$ leads to at least one symbol change in $\{X_t(k+1)\}$. Consequently, we should have $D_{k+1} \geq D_k$, a fact that reveals the monotonic character of simple HOC. In finite data records, what we really have is the sure inequality $D_{k+1} \geq D_k - 1$ [127]. In addition, as k increases, the discrimination power of simple HOC diminishes, because different processes yield almost the same D_k [127]. A value of $K = 20$ usually suffices for effective discrimination purposes [127].

2.6.3   HOC Discrimination Tools

HOC-based discrimination between signals consists of the process of determining from HOC sequences their degree of similarity. Sometimes, instead of comparing directly the two signals, it is useful to measure their distance from a given reference signal. In many respects, white Gaussian noise is the most convenient, for which the expected HOC are given by [127]

$$E\{D_k\} = (N - 1)\left\{ \frac{1}{2} + \frac{1}{\pi}\sin^{-1}\left(\frac{k-1}{k}\right) \right\}. \tag{2.76}$$

The limits for which D_k falls within with a probability of approximately 95% for each k are [127]

$$\pm 1.96(N - 1)^{1/2}\left\{ \frac{1}{4} - \left[\frac{1}{\pi}\sin^{-1}\left(\frac{k-1}{k}\right)\right]^2 \right\}^{1/2}. \tag{2.77}$$

With (2.76) and (2.77), a *white noise test* (WNT) could be constructed to examine whether a time series oscillates as Gaussian white noise. Based on WNT, the hypothesis of white noise should be rejected when at least one D_k, $k = 1, 2, \ldots, K$, for some K, falls outside the limits of (2.76) [127].

Using WNT, each time series is tested against the white Gaussian noise. By superimposing the results from both tests in the HOC domain, the two time series are actually compared to each other. This indirect comparison could reveal the optimum order, k_{op}, of HOC that results in the maximum discrimination of the two time series, corresponding to the maximum distance between their HOC sequences.

Based on the results of WNT, scatter plots of HOC of k_{op} order versus HOC of other (usually neighboring) orders can reveal the presence of two classes (clusters). If the classes do not overlap in the scatter plots, linear discrimination could be achieved and a simple decision rule of the form

$$aD_{k_{op} \pm n} + bD_{k_{op}} + c = 0, \ n = 1 \text{ or } 2 \text{ or } 3; \ a,b,c = \text{constants}, \quad (2.78)$$

could be constructed in deciding upon class membership. This also means that the HOC sequences of the two classes do not overlap in the WNT, at least for one value of k. Otherwise, linear discrimination is not feasible and the classes overlap in the scatter plots.

To examine the possibility of potential overlapping between the two groups, the mean (MN) and the standard deviation (σ) of the distance from cluster center, alongside the standard error of the MN (s), both for the two classes in the HOC scatter plot, are estimated. Consequently, the borderline of the two clusters is defined as the perpendicular line at the $\mathbf{C}(l_1, l_2)$ point of their center distance, with the $\mathbf{C}$ vector satisfying the following equations:

$$\frac{\text{dis}(\mathbf{C}, \mathbf{c}_1)}{\text{MN}_1 + \sigma_1 + s_1} = \frac{\text{dis}(\mathbf{C}, \mathbf{c}_2)}{\text{MN}_2 + \sigma_2 + s_2} \quad (2.79)$$

$$\text{dis}(\mathbf{C}, \mathbf{c}_1) + \text{dis}(\mathbf{C}, \mathbf{c}_2) = \text{dis}(\mathbf{c}_1, \mathbf{c}_2), \quad (2.80)$$

where $\text{dis}(\mathbf{a}_1, \mathbf{a}_2)$ denotes the Euclidean distance between vectors $\mathbf{a}_1$ and $\mathbf{a}_2$, and $\mathbf{c}_i$, MN_i, σ_i, s_i, $i = 1, 2$, denote the estimated center vector, the mean, the standard deviation, and the standard error of the mean for the two clusters, respectively.

For an extended coverage of the HOC-based analysis of time series, the reader is encouraged to consult the book by Kedem [127] and its references.

2.6.4   HOC Analysis of DAS

HOC have been used to perform a discrimination analysis of DAS, especially FC, CC, and SQ [128]. In fact, HOC assesses the changes in oscillatory pattern of DAS according to their type, by estimating the relevant HOC sequence. Based on the connection between filtering and zero crossings, the proposed method constitutes a possible tool for DAS discrimination analysis (namely,

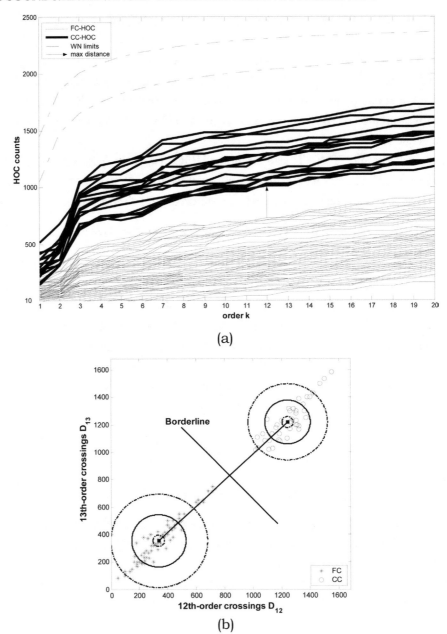

(a)

(b)

FIGURE 2.9: (a) WNT for FC and CC with k order ranging from 1 to 20. (b) Scatter plot of 12th-order (D_{12}) vs. 13th-order (D_{13}) crossings of FC (stars) and CC (circles). The ■ symbol denotes the center of each group (FC and CC), whereas the borderline is the estimated linear border that clearly discriminates the two groups. The solid and dash-dotted circles correspond to MN and $[\pm(\sigma + s)]$ of the distance from cluster center, respectively, both for FC and CC [128].

HOC-DA), which is quite attractive because of its simplicity in the evaluation of the HOC sequence and its ability to provide simple discrimination criteria between the examined classes, when it is feasible.

An example of HOC-DA is given in Figure 2.9a, which illustrates the results from WNT for FC and CC. The estimated HOC sequences of FC (plain solid lines) and CC (bold solid lines) are directly compared to the white Gaussian noise limits (dash-dotted lines) and indirectly to each other. The HOC order k ranged from 1 to 20 and its optimum value was found k_{op}^{FC-CC} = 2 because there, the distance (denoted by an arrow) between the topmost HOC line of FC and the bottommost HOC line of CC is maximized. This implies that scatter plots of the k_{op}^{FC-CC} -order crossings could provide a clear (linear) discrimination between FC and CC. This is true from the HOC scatter plot of Figure 2.9b, where the HOC pair (D_{12}, D_{13}) from the time series of FC (stars) and CC (circles) is depicted. Clearly, the two processes give rise to distinctly different clusters of pairs (D_{12}, D_{13}), which can be linearly discriminated. In this scatter plot, the coordinates of the two cluster centers were found equal to c_1^{FC} (338.8, 354.5) and c_1^{CC} (1242.6, 1218.7) with a distance of dis(c_1^{FC}, c_1^{CC}) = 1251.2, and are denoted as solid squares connected with solid line, respectively. Furthermore, the estimated MN_1^{FC} = 190.7; σ_1^{FC} = 133.9; s_1^{FC} = 15.6 and MN_1^{CC} = 158.9; σ_1^{CC} = 103.2; s_1^{CC} = 16.1 are also depicted as solid circles (MN) and dash-dotted ones $[\pm(\sigma + s)]$, both for FC and CC clusters. The $\mathbf{C}^{FC-CC}$ vector was derived by (2.79) and (2.80), using as $\mathbf{c}_1 = \mathbf{c}_1^{FC}$, $\mathbf{c}_2 = \mathbf{c}_1^{CC}$, $MN_1 = MN_1^{FC}$, $\sigma_1 = \sigma_1^{FC}$, $s_1 = s_1^{FC}$, $MN_2 = MN_1^{CC}$, $\sigma_2 = \sigma_1^{CC}$, $s_2 = s_1^{CC}$ and its coordinates were found equal to $\mathbf{C}^{FC-CC}$ (836.1, 830), defining the borderline shown in Figure 2.9b.

The above results show that FC and CC are linearly discriminated and justify the selection of "FC range" and "CC range" that provide discrimination with an accuracy of 100% (SE = 100%, SP = 100%) between FC and CC in the HOC domain as

$$r_{100\%}^{FC-CC} = \left\{ [D_{12}(40 \div 800), D_{13}(40 \div 800)]^{FC}, \ [D_{12}(1000 \div 1600), D_{13}(1000 \div 1600)]^{CC} \right\}$$

Further examples of the application of the HOC-DA to {FC, SQ } and {CC, SQ } DAS pairs are included in [128]. The described analysis show the ability of the HOC-based discrimination tool to efficiently reveal the differences in the oscillatory patterns of the DAS types that occur when higher (>1) order crossings are used and sets a different perspective in the time-domain analysis of LS.

2.7 EMPIRICAL MODE DECOMPOSITION

2.7.1 Epitomized Rationale

The HOC analysis presented in the previous section deals mainly with (in a broad sense) stationary signals. To accommodate the nonstationarity of the processes, the zero-crossing rate could be estimated either within a moving data window or adaptively, giving more weight to more current data as a means for summarizing a time dependent oscillation pattern. Nevertheless, a more efficient approach lies in the field of empirical mode decomposition (EMD), proposed by Huang et al. [129] in 1998. EMD decomposes the signal into components with well-defined instantaneous frequency. Each characteristic oscillatory mode extracted, namely, intrinsic mode function (IMF), is symmetric and has a unique local frequency, and different IMFs do not exhibit the same frequency at the same time [129].

The EMD method is necessary to deal with both nonstationary and nonlinear data and, contrary to almost all the previous methods, EMD is intuitive—that is, the basis of the expansion is generated in a direct, a posteriori, and adaptive manner, derived from the data [129]. The main idea behind EMD is that all data consist of different simple intrinsic modes of oscillations, represented by the IMFs. An IMF represents a simple oscillatory mode as a counterpart of the simple harmonic function, yet it allows amplitude and frequency modulation; thus, it is much more general. The EMD method considers the signals at their local oscillation scale, subtracts the faster oscillation, and iterates the residual.

EMD can also be viewed as an expansion of the data in terms of IMFs. Then, these IMFs, based on and derived from the data, can serve as the basis of that expansion, which can be linear or nonlinear as dictated by the data, and it is complete and almost orthogonal. Most important of all, it is adaptive, and, therefore, highly efficient.

Representation of LS in the EMD domain reveals their intrinsic characteristics, because these are reflected in the IMFs, and provides a new analysis space where detailed exploitation of LS diagnostic information is feasible.

2.7.2   EMD Description

EMD is formed through the estimation of IMFs. According to Huang et al. [129], an IMF satisfies two conditions:

- In the whole dataset, the number of extrema and the number of zero crossings must either equal or differ at most by one.
- At any point, the mean value of the envelope defined by the local maxima and the envelope defined by the local minima is zero.

By virtue of the IMF definition, the EMD procedure for a given signal $x(t)$ can be summarized as follows [129]:

- Identify the successive extrema of $x(t)$ based on the sign alterations across the derivative of $x(t)$.

- Extract the upper and lower envelopes by interpolation, that is, the local maxima (minima) are connected by a cubic spline interpolation to produce the upper (lower) envelope. These envelopes should cover all the data between them.
- Compute the average of upper and lower envelopes, $m_1(t)$.
- Calculate the first component $h_1(t) = x(t) - m_1(t)$.
- Ideally, $h_1(t)$ should be an IMF. In reality, however, overshoots and undershoots are common, which also generate new extrema and sift or exaggerate the existing ones [129]. To correct this, the sifting process has to be repeated as many times as is required to reduce the extracted signal as an IMF. To this end, treat $h_1(t)$ as a new set of data, and repeat first to fourth steps up to k times (e.g., $k = 7$) until $h_{1k}(t)$ becomes a true IMF. Then set $c_1(t) = h_{1k}(t)$. Overall, $c_1(t)$ should contain the finest scale or the shortest period component of the signal.
- Obtain the residue $r_1(t) = x(t) - c_1(t)$
- Treat $r_1(t)$ as a new set of data and repeat first to sixth steps up to N times until the residue $r_N(t)$ becomes a constant, a monotonic function, or a function with only one cycle from which no more IMFs can be extracted. Note that even for data with zero mean, $r_N(t)$ still can differ from zero.
- Finally,

$$x(t) = \sum_{i=1}^{N} c_i(t) + r_N(t),\qquad(2.81)$$

where $c_i(t)$ is the ith IMF and $r_N(t)$ is the final residue.

The above procedure results in a decomposition of the data into N-empirical modes, and a residue $r_N(t)$, which can be either a monotonic function or a single cycle. It is noteworthy that, in order to apply the EMD method, there is no need for a mean or zero reference; EMD only needs the locations of the local extrema to generate the zero reference for each component (except for the residue) through the shifting process.

A useful characteristic of (2.81) is the potential of filtering that it provides. Indeed, using the IMF components, a time-space filtering can be devised simply by selecting a specific range of them in the reconstruction procedure [e.g., in (2.81), for high-pass filtering: $i = 1{:}k$, $k < N$, or for band pass one: $i = b{:}k$, $1 < b$, $k < N$). This time-space filtering has the advantage that its results preserve the full nonlinearity and nonstationarity in the physical space [130].

2.7.3   EMD Considerations and Extensions

As described above, the process of finding IMFs is indeed like sifting: to separate the finest local mode from the data first based only on the characteristic timescale. The sifting process, however, has two effects: (1) to eliminate riding waves and (b) to smooth uneven amplitudes. Although the first condition is absolutely necessary for the instantaneous frequency to be meaningful, the second condition is also necessary in case the neighboring wave amplitudes have too large a disparity. Unfortunately, the second effect, when carried to the extreme, could obliterate the physically meaningful amplitude fluctuations. Therefore, the sifting process should be applied with care, because carrying the process to an extreme could make the resulting IMF a pure frequency-modulated signal of constant amplitude.

To guarantee that the IMF components retain enough physical sense of both amplitude and frequency modulations, we have to determine a criterion for the sifting process to stop. This can be accomplished by limiting the size of the standard deviation, SD, computed from the two consecutive sifting results [131, 132]. Historically, two different criteria have been proposed. The first one was introduced by Huang et al. [129], determined by using a Cauchy type of convergence test. Specifically, the test requires the normalized squared difference between two successive sifting operations defined as

$$\mathrm{SD}_k = \frac{\sum_{t=0}^{T} \left| h_{1(k-1)}(t) - h_{1k}(t) \right|^2}{\sum_{t=0}^{T} h_{1(k-1)}^2(t)},$$

(2.82)

to be small. If this squared difference SD_k is smaller than a predetermined value, the sifting process will be stopped. This definition seems to be rigorous, but it is very difficult to implement in practice. This criterion does not depend on the definition of the IMFs. The squared difference might be small, but nothing guarantees that the function will have the same numbers of zero crossings and extrema, for example.

These shortcomings prompted Huang et al. [133, 134] to propose a second criterion based on the agreement of the number of zero crossings and extrema. Specifically, an S number is preselected. The sifting process will stop only after S consecutive times, when the numbers of zero crossings and extrema stay the same and are equal or differ at most by 1. This second choice has its own difficulty: how to select the S number. Obviously, any selection is ad hoc, and a rigorous justification is needed. In a recent study of this open-ended sifting, Huang et al. [134] used the many possible choices of S numbers to form an ensemble of IMF sets, from which an ensemble mean and confidence were derived. Furthermore, through comparisons of the individual sets with the mean, Huang et al. [134] established an empirical guide. For the optimal siftings, the range of S number should be set between 4 and 8.

Ensemble EMD. One of the possible drawbacks of EMD is the occasional appearance of *mode mixing*, which is defined as a single IMF either consisting of signals of widely disparate scales, or a signal of a similar scale residing in different IMF components. Mode mixing is a consequence of signal intermittency. Recently, a new sifting approach was proposed, Ensemble EMD (EEMD) [135–137], which successfully deals with the mode mixing problem. This new approach consists of sifting an ensemble of white noise-added signal and treats the mean as the final true result. Finite, not infinitesimal, amplitude white noise is necessary to force the ensemble to exhaust all possible solutions in the sifting process, thus making the different scale signals to collate in the proper IMFs dictated by the dyadic filter banks. Because EMD is a time-space analysis method, the white noise is averaged out with sufficient number of trials; the only persistent part survives the averaging process is the signal, which is then treated as the true and more physical meaningful answer. The effect of the added white noise is to provide a uniform reference frame in the time–frequency space; therefore, the added noise collates the portion of the signal of comparable scale in one IMF. With this ensemble mean, one can separate scales naturally without any a priori subjective criterion selection as in the intermittence test for the original EMD algorithm. This new approach uses the full advantage of the statistical characteristics of white noise to perturb the signal in its true solution neighborhood, and to cancel itself out after serving its purpose; therefore, it represents a substantial improvement over the original EMD.

Complementary ensemble EMD. An extension to the EEMD was just recently proposed by Yeh et al. [138], with the purpose of removing the residue of added white noises. They modified the EEMD method by using the complementary sets of added white noises to remove residue of added white noises. This modified EEMD, complementary EEMD (CEEMD), was also developed as a new technique to overcome intermittence, hence, mode mixing.

CEEMD extends the concept of EEMD method by generating two sets of averaged IMFs, that is, averaged IMFs with positive and negative residues of added white noises. In CEEMD, a number (e.g., 20) of arbitrary white noises are chosen for added white noises, because a single white noise cannot solve all intermittent signals. Two mixtures of the original signal and added white noise are then produced by the following equation:

$$\begin{bmatrix} M_1 \\ M_2 \end{bmatrix} = \begin{bmatrix} 1 & 1 \\ 1 & -1 \end{bmatrix} \begin{bmatrix} S \\ N \end{bmatrix} \qquad (2.83)$$

where S is the original signal, N is the added white noise, M_1 is the positive mixture, and M_2 is the negative mixture.

The averaged IMFs decomposed from positive mixtures are the averaged IMFs with positive residue of added white noises. Similarly, the averaged IMFs with negative residue of added white noises are obtained. Then, the averaged IMFs without added residue of white noises can be derived by the following equation

$$\begin{bmatrix} \text{IMF}_S \\ \text{IMF}_N \end{bmatrix} = \text{inv} \begin{bmatrix} 1 & 1 \\ 1 & -1 \end{bmatrix} \begin{bmatrix} \text{IMF}_p \\ \text{IMF}_n \end{bmatrix}, \qquad\qquad (2.84)$$

where IMF_p is the averaged IMF with positive residue of added white noises; IMF_n is the averaged IMF with negative residue of added white noises; IMF_S is the averaged IMF without the residue of added white noises; IMF_N is the part of averaged IMF contributed by the added white noises. The averaged IMF without the residue of added white noises is the final result of CEEMD [138].

For the signal reconstruction process, only significant IMFs are used, defined through Monte Carlo verification. In particular, for a normalized white noise, the distribution of energy density-correspondent averaged period of IMFs should locate around the diagonal of Monte Carlo plot [135, 139]. For a normalized signal with significant components, the relationship between the energy density and averaged period of a significant IMF should locate on the area above the diagonal. Hence, based on this, significant IMFs are identified [138].

2.7.4   EMD Crackles Analysis

EMD was applied to describe crackles behavior [140] according to the characteristics of the corresponding IMFs. The main findings of the study [140] were:

- Oscillatory information of crackles embedded in basic inspiratory noise are distributed on several IMFs, starting FC information at IMF_2 and CC at IMF_3, yet the IMF that contains the ending information is not clear.
- Oscillatory information of basic inspiratory sounds starts at IMF_3, but again, the last IMF containing basic respiratory sound information is not clear.
- Depending on the SNR, FC are easier to discriminate from basic respiratory sound than CC.
- Overlapped crackles appear even at IMF_1, but individual components are not differentiated.
- Combined but time-separated FC and CC have no effect on the oscillatory information for individual events.

As Villalobos et al. [140] have stated, the EMD technique improved the visual identification of crackles embedded in respiratory sound. Although crackles and respiratory sound were mixed, the authors were able to identify their respective oscillations at different IMFs. For example, FC were more easily observed in IMF_2, whereas CC were associated to oscillations in IMF_3, probably due to their spectral content differences. Such distinction, however, depends on the intensity of the background noise and how this noise changes the crackle morphology. Because crackles information is propagated to more than one IMF, combining information from different IMFs could help to identify even low amplitude FC and CC and discard artifacts. Therefore, EMD analysis may assist, in addition to crackles counting, in determining the crackles type. EMD analysis is advantageous because it generates oscillations that are associated with crackle locations, and it could separate individual FC that overlapped by a small amount.

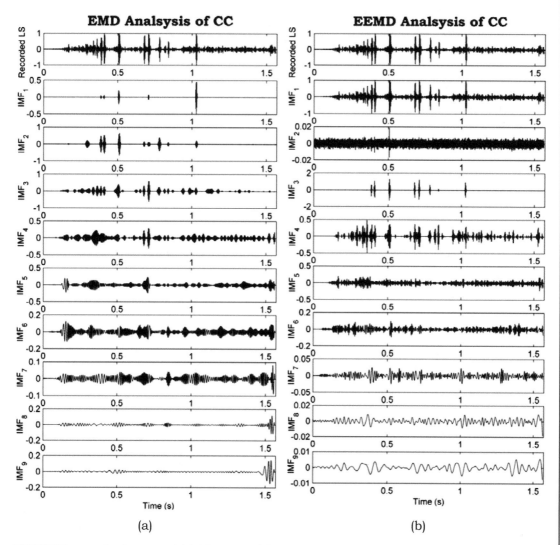

FIGURE 2.10: Application of (a) EMD and (b) EEMD to LS containing CC superimposed on vesicular LS (see first part of Figure 1.1c). In both cases, the recorded LS signal along with the first nine IMFs is depicted, respectively.

An example of the application of EMD and EEMD to recorded CC is illustrated in Figure 2.10a and b, respectively. In particular, the recorded LS signal along with the first nine IMFs is depicted, showing the distribution of the vesicular sound and CC waveforms across the decomposition levels. When comparing the morphology of the IMFs between the two decompositions, some differences can be identified. In particular, IMF_1 clearly differs from the original signal in the case of EMD (Figure 2.10a), whereas in the case of EEMD (Figure 2.10b) it exhibits strong similarity with the original signal. CC are more isolated in the first four EMD-IMFs, whereas CC dominate in the third and fourth EEMD-IMFs. Clearly, $EEMD\text{-}IMF_2$ corresponds to the additive noise used in the EEMD realization. In general, EEMD-IMFs seem to better represent CC than the EMD-IMFs, despite the domination of noise in some IMFs, because EEMD reduces the mode mixing effect (e.g., at IMF_3 and IMF_4 level).

EMD analysis has also been used in a DAS denoising process, as will be thoroughly described in the next chapter.

2.8 FRACTAL DIMENSION–LACUNARITY ANALYSIS

2.8.1 Epitomized Rationale

Fractal dimension. The term "fractal dimension" (FD) can more generally refer to any of the dimensions commonly used for fractals characterization (e.g., capacity dimension, correlation dimension, information dimension, Lyapunov dimension, Minkowski–Bouligand dimension) [141]. In other words, FD is a measure of how "complicated" a self-similar figure is. To this end, FD can be considered as a relative measure of the number of basic building blocks that form a pattern [142]. Consequently, FD could reflect the signal complexity in the time domain. This complexity could vary with sudden occurrence of transient signals, such as explosive LS.

As Gnitecki and Moussavi [62] note, LS will possess valid FD values, based on their morphological properties. Most obviously, the signals do not self-cross. In physical systems, the property of self-similarity in pure fractal objects is not strict but is probabilistic, and there are minimum and maximum scaling limits [143]. In reality, an object occurring in nature, such as a physiological signal, that exactly duplicates itself over several scales does not exist. Thus, the fractality of LS is in the self-affine sense [62]. Furthermore, LSs exhibit clear quasiperiodicity because they emerge from natural biological processes, such as breathing, which implies that they are not purely random [143]. Consequently, FD analysis can shed light on LS from a functional point of view.

Lacunarity. Laennec [144], in his attempt to develop a metalanguage of sound (as reported in his magnum opus *Treatise*), defined a set of descriptions for the shape and texture of sounds that was independent of subjective experience (i.e., independently verifiable). By using the texture of LS in this codification, Laennec showed the importance of this sound property in the hydraulic hermeneutics of mediated auscultation. Following this pathway, LS texture could be approached through the concept of *lacunarity*. The latter was originally developed to describe a property of fractals [145–148], and to discriminate textures and natural surfaces that share the same FD.

Gefen et al. [147] define lacunarity as the deviation of a fractal from translational invariance. Translational invariance can also be a property of nonfractal sets [148], and it is highly scale-dependent; sets that are heterogeneous at small scales can be quite homogeneous when examined at larger scales or vice versa [149]. From this perspective, lacunarity can be considered a scale-dependent measure of heterogeneity or texture of an object, whether or not it is fractal [148]. Lacunarity features can be used to assess the largeness of gaps or holes of one- (signals) or two-dimensional sets (images). It can describe the complex intermingling of the shapes and the distribution of gaps within a set. A set with low lacunarity is homogeneous and transitional invariant, whereas one with high lacunarity has gaps distributed across a broad range of sizes [150].

2.8.2   FD Estimation

There are two principal approaches to estimate the FD of a time series, one that operates directly on the waveform and one that operates in a reconstructed state space [151]. In the first approach, FD is derived from an operation directly on the signal and not on any state space. This means that the data series does not have to be embedded into higher dimensional space for the FD estimation; hence, this type of FD estimation has fast computational implementation. In the second approach, the waveform is regarded as a planer set in R^2, where it is considered a geometric object. A typical time series has a dimension somewhere in between 1 and 2, as it is more complicated than a line but never covers the whole two-dimensional space.

There are several estimation techniques of FD [152], a few of which are *box counting dimension* (based on counting how many d-dimensional hypercubes with side length ε are needed to cover an attractor), *information dimension* (extended version of the box counting dimension by weighting its count by measuring how much of the attractor is contained within each hypercube), *correlation dimension* (a measure of how the number of neighbors in the reconstructed state space varies with decreasing neighborhood sizes ε), *Katz FD* [153], *Sevcik FD* [154], *variance FD* [143]. The latter three are described more analytically below.

Katz FD. According to Katz [153], the FD of a curve defined by a sequence of N points is estimated by

$$\mathrm{FD}^{\mathrm{K}} = \frac{\log_{10}(n_s)}{\log_{10}\left(\dfrac{d}{L_c}\right) + \log_{10}(n_s)}, \tag{2.85}$$

where K denotes Katz's definition of FD; L_c is the total length of the curve, realized as the sum of distances between successive points, that is,

$$L_c = \sum_{i=1}^{N-1} \text{dist}(i, i+1), \tag{2.86}$$

where $\text{dist}(i, j)$ is the distance between the i and j points of the curve; d is the diameter estimated as

$$d = \max[\text{dist}(i, j)]; \ i \neq j; \ i, j \in [1, N]; \tag{2.87}$$

for curves that do not cross themselves, usually, d diameter is estimated as the distance between the first point of the sequence and the point of the sequence that provides the farthest distance [153], that is,

$$d = \max[\text{dist}(1, i)]; \ i \in [2, N]; \tag{2.88}$$

and n_s is the number of steps in the curve, defined as

$$n_s = \frac{L_c}{a}, \tag{2.89}$$

where $\bar{a}$ denotes the average step, that is, the average distance between successive points. In this way, a general unit or "yardstick" is formed that eliminates the dependence of the FD estimates derived by (2.85) on the measurement units used [153].

Sevcik FD. Sevcik [154] uses the following definition for the FD of an N-sample curve:

$$\text{FD}^S = 1 + \frac{\ln(L_c)}{\ln[2 \cdot (N-1)]}, \tag{2.90}$$

where S denotes Sevcik's definition of FD and L_c is defined as in (2.86). Before applying (2.90), Sevcik proposes, for convenience, linear transformations of the waveform, in order to transform it into a normalized space where all axes are equal [154]. He proposes normalization of every point in the abscissa as:

$$x_i^* = \frac{x_i}{x_{\max}}, \ i = 1, \dots, N, \tag{2.91}$$

where x_i, $i = 1, \dots, N$, are the original values of the abscissa and $x_{\max} = \max(x_i)$, and ordinate normalization as:

$$y_i^* = \frac{y_i - y_{\min}}{y_{\max} - y_{\min}}, \ i = 1, \dots, N, \tag{2.92}$$

where y_i, $i = 1, \dots, N$, are the original values of the ordinate, $y_{\max} = \max(y_i)$ and $y_{\min} = \min(y_i)$.

Variance FD. The variance FD (VFD) of a $s(t)$ signal (i.e., LS) is estimated via the Hurst exponent given by

$$H = \lim_{\Delta t \to 0} \left[\frac{\frac{1}{2} \log\{\mathrm{var}\,[(\Delta s)_{\Delta t}]\}}{\log(\Delta t)} \right], \tag{2.93}$$

which is a measure of signal smoothness. In (2.92), $(\Delta s)_{\Delta t}$ denotes $s(t_2) - s(t_1)$ and $\Delta t = |t_2 - t_1|$. Using (2.92), VFD is given by

$$\mathrm{VFD} = E + 1 - H, \tag{2.94}$$

where E corresponds to the Euclidean dimension, which equals one for a 1-D time series; hence, (2.93) can be rewritten as

$$\mathrm{VFD} = 2 - H. \tag{2.95}$$

In fact, (2.94) implies a power-law relationship between the variance of the amplitude increment of $s(t)$, which is produced by a dynamical process, over a time increment Δt. This power-law is of the form

$$\mathrm{var}\,[(\Delta s)_{\Delta t}] \propto \Delta t^{2H}. \tag{2.96}$$

The choice of the time increment depends on the aim of the application; unit time increment is preferred when separating a signal from noise, whereas a dyadic is preferred for separating different components within the signal [143].

2.8.3   Lacunarity Estimation

There are many algorithms to define, quantify, and calculate lacunarity [145, 147]. In recent years, the most popular methods all seem to be based on the use of the intuitively clear and computationally simple "gliding box" method of Allain and Cloitre [148]. Stemming from the latter, Plotnick et al. [149] extended the concept of lacunarity to describe the spatial distribution of a real dataset even if it is not an ideal fractal set. Furthermore, Dong [155] and Du and Yeo [150] proposed two lacunarity estimating methods for image texture analysis and segmentation, respectively. The "gliding box" method of Allain and Cloitre [148] consists of the following main steps.

Consider a set of data with length M; at first, a gliding box of length $r(\leq M)$ is placed at the origin of the set and the number of occupied sites within the box, that is, locations with existence of information, termed as "box mass" s, are counted. Next, the box is moved to the right one space along the set and the box mass is counted again. This process is repeated over the entire set, producing a frequency distribution of the box masses $n(s,r)$, which is then converted into probability distribution $Q(s,r)$ by dividing by the total number of boxes $N(r)$ of size r. Then, the first and second moments of $Q(s,r)$ are estimated by

$$Z_1 = \sum_s s Q(s,r), \tag{2.97}$$

$$Z_2 = \sum_s s^2 Q(s,r). \tag{2.98}$$

Using (2.97) and (2.98), the lacunarity for size r is defined as

$$\Lambda(r) = \frac{Z_2^2}{Z_1^2}. \tag{2.99}$$

The calculation of (2.99) is repeated over a range of box sizes, ranging from $r = 1$ to some fraction of the length of the analyzed data M. Considering the form of (2.97) and (2.98), it is clear that they can be rewritten in the form of

$$Z_1 = \overline{s(r)}, \tag{2.100}$$

$$Z_2 = \sigma_s^2(r) + \left(\overline{s(r)} \right)^2, \tag{2.101}$$

with $\overline{s(r)}$ and σ_s^2 denoting the mean and the variance of the number of sites per box, respectively. Consequently, (2.99) could be written as

$$\Lambda(r) = \frac{\sigma_s^2(r)}{\left(\overline{s(r)} \right)^2} + 1. \tag{2.102}$$

Based on (2.102), it is clear that the lacunarity statistic is a dimensionless representation of the variance to mean ratio. The representation of lacunarity in (2.102) reveals its texture relationship. In particular, sparse regions have low means of the box mass; in other words, they appear to be less homogeneous than dense regions. Consequently, they will have higher lacunarity values than dense regions while counted with the same box size. For the same region, large boxes will generally be more homogeneous than small ones, so the lacunarity value of the former case is lower than that of the latter. Before the size of the boxes reaches the scale of randomness of the dataset, which is the scale at which gaps in the dataset appear to be almost homogeneous, significant variation in lacunarity values can be observed. When the variance of the gaps between occupied cells approaches zero, lacunarity values approach one [150].

Original applications of lacunarity analysis have used binary datasets only [147, 148]. However, as Plotnick et al. [149] denote, the use of quantitative (instead of binary) data is analogous to beginning the analysis at a coarser level of resolution. Lacunarity can thus be calculated by using the sum (or integral) of the distribution in a box of size r.

2.8.4   FD Analysis of LS

An analysis of LS with FD is presented in the work of Gnitecki and Moussavi [62], who examined the fractality of LS from normal subjects reflected in the estimated values of FD using FD^K, FD^S, and VFD, defined in (2.85), (2.90), and (2.95), respectively. LS were sequestered corresponding to 85–100% of the maximum flow per inspiratory breath. Three sliding window sizes were used for each FD calculation across the signals (50, 100, and 200 ms, respectively), shifted forward by 25 ms.

In [62], apparent similarities, in terms of both morphology and overall variability, between three scalings of LS were found. Overall, both VFD and FD^K showed an increased with airflow, whereas FD^S appeared to be unaffected by change in flow and hence in LS intensity. From the characteristics of the FD analysis of the examined normal LS, a fractal character was identified; this finding, however, was not justified for the case of abnormal LS.

FD was also used in the analysis of DAS for their detection and isolation from the vesicular sound; this is thoroughly described in the next chapter.

2.8.5   Lacunarity Analysis of DAS

The texture-based approach of LS using lacunarity analysis is a fairly new concept. Application of lacunarity analysis to DAS classification has been recently examined in [83]. In this work, denoised signals from each DAS category, that is, FC, CC, and SQ, formed data series that correspond to the three textures, respectively, as illustrated in Figure 2.11(i). Thirty-two sections of the same size W from every texture (data series) were randomly chosen. Next, from these 32 sections, 24 sections (75%) were used to estimate the lacunarity value of the texture, that is, these 24 sections were used to train a classifier. The remaining eight sections (25%) were used to test the classification performance. This sequence was repeated 200 times, and classification accuracy was averaged out. By repeating the classification experiment many times with different (random) training and test sections, robust estimate of the classifier performance and the texture (i.e., DBS) description ability of the proposed lacunarity-based approach were obtained. The classification process used in [83] was the discriminant analysis [156, 157]. The classification was based on the Mahalanobis method, which uses Mahalanobis distances with stratify covariance estimates [156, 157].

Figure 2.11(ii) depicts the estimated normalized lacunarity $\Lambda(r)$ [see (2.99)] for the FC–CC–SQ comparison group for the training and testing dataset, respectively, versus the box size r when

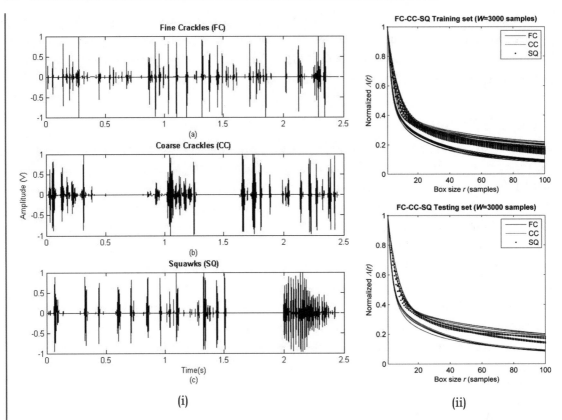

FIGURE 2.11: Lacunarity analysis of DAS. (i) The three "texture-series" corresponding to FC, CC, and SQ. (ii) Estimated normalized lacunarity $\Lambda(r)$ for the training (top) and testing (bottom) set with $W = 3000$ samples ($\Lambda(r)$ that corresponds to FC, CC, and SQ is plotted with solid, dashed, and dotted line, respectively) [83].

$W = 3000$ samples (sampling at 2.5 kHz). In Figure 2.11(ii), it is clear that curves are grouped in a manner that makes texture separation feasible. In particular, both in the training and testing sets, FC-related lacunarity curves are grouped in a markedly distant manner from the lacunarity curves related to other two textures (CC and SQ), almost independently of the box size, whereas CC- and SQ-related lacunarity curves show a more neighborly behavior that makes their separation dependent on the value of box size r.

Results shown in Figure 2.11(ii) indicate that the behavior of the normalized $\Lambda(r)$ resembles that of the hyperbola. In particular, during the transition from CC to SQ and further to FC cases, the concavity of the hyperbola function changes, corresponding to a transition from higher to lower lacunarity, respectively. This change in the concavity of the hyperbola function seems to be the basic parameter that contributes to the successful categorization among DBS types using lacunarity analysis. The higher values of lacunarity observed in CC compared to those in FC are due to the difference in their reproduction. CC behave more randomly than FC over subsequent breaths [12]; hence, they create sparser regions [see Figure 2.11(i), b] that have low means of the box mass, reduce their homogeneity, and result in higher lacunarity values than those for the FC regions, while counted with the same box size. An intermediate situation is seen in the case of SQ, where the combination of FC with a short wheeze results in less sparse regions than CC, yet more than FC [see Figure 2.11(i), c), creating smaller lacunarity values than CC, but higher than those in FC [see Figure 2.11(ii)].

When applied to two datasets of DAS, lacunarity analysis exhibited high classification performance, yielding results with a combined mean classification accuracy of more than 99.6% for all comparison groups (FC–CC, FC–SQ, CC–SQ, FC–CC–SQ) [83]. The randomized training and testing sections used in the evaluation of lacunarity analysis augurs a promising performance of LAC analysis under different case scenarios. The simplicity of LAC analysis facilitates the customization of its computational environment under a real-time context, resulting in an easily implemented and user-friendly processing tool of DAS.

. . . . .

CHAPTER 3

Denoising Techniques

3.1 OVERVIEW

Noise contamination of the signal of interest is an ordinary effect in signals acquired from pragmatic systems, such as the human body. When focusing on lung sounds (LS), this is clearly evident, as different types of noises coexist with the useful LS; heart sounds, respiratory muscle sounds, friction rubs, ambient noise sources are some primary examples. Because physicians include LS interpretation, alongside other symptomatic indicators, in their process of diagnosing a respiratory pathology, denoising of useful information could contribute to more accurate and objective appreciation of LS information.

In this chapter, a series of denoising techniques applied to LS enhancement are described as indicative examples. It should be noted that the concept of denoising adapts to the specific problem examined, as, for instance, DAS denoising refers to their detection and extraction from the vesicular sound (VS), whereas heart sound cancellation focuses on leaving most unaffected the remaining LS after the efficient removal of heart sounds.

The techniques described here refer directly or indirectly to the new domains of LS representation presented in the previous chapter; hence, the latter could serve as a background basis for the methodologies adopted in the denoising techniques presented in this chapter.

3.2 WAVELET-BASED DENOISING

Two LS denoising techniques based on wavelets are presented: continuous wavelet transform–wheeze episode detector (CWT-WED) [158] and wavelet transform-based stationary–nonstationary (WTST-NST) filter [46]. CWT-WED refers to the use of CWT in wheeze detection and extraction from breathing sound, whereas the WTST-NST filter provides efficient separation of discontinuous adventitious sounds (DAS) from VS (background-type noise).

3.2.1 CWT-WED

In the CWT-WED algorithm, the focus is placed on the automatic location and identification of wheezing episodes during breathing. In particular, CWT (see (2.47)) is used to form a wheezing episode detector taking into account the beneficial properties of CWT representation over the Fourier one, because, unlike the latter, CWT provides useful information on the coherent nature of localized features within the signal, such as wheezing episodes.

CWT-WED is structured in a twofold manner. In particular, it not only focuses in identifying the true location of wheezes within the breathing cycle, it also aims at separating the wheeze signal from the acquired lung sound. In this way, the diagnostic character of the wheeze is unveiled, because its contamination due to the superimposition of the breathing sound is circumvented. As a result, CWT-WED acts both as an identification and denoising tool.

Wheeze identification. The N-sample acquired lung sound signal, $x(k)$, can be considered as the sum of an envelop "confined" sinusoidal signal of interest, $s(t)$, that is, wheezes, and the breathing sound, $n(k)$ (such as background noise); hence, it can be written as

$$x(k) = s(k) + n(k), \; k = 1, \ldots, N. \tag{3.1}$$

CWT is realized using the real Morlet wavelet [102], defined as

$$\psi(t) = \sqrt{\pi f_b} \exp\left(-t^2 / f_b\right) \cos\left(2\pi f_0 t\right), \tag{3.2}$$

where f_b is the bandwidth parameter that controls the Gaussian envelope, which "confines" the sinusoidal waveform with a center frequency of f_0. By taking the square of the modulus of the resulted CWT, the scalogram is estimated as

$$S_c(a, b) = |W_s(a, b)|^2 \tag{3.3}$$

which is analogous to the spectrogram [102], and highlights the location and scale of dominant energetic features within the signal. The increased coherency between the signal of interest, $s(t)$ (i.e., wheezes), and the analyzing wavelet, unlike the breathing sound, $n(k)$w, is reflected in the scalogram by presenting distinct isolated peaks. Consequently, unlike the spectrogram, the scalogram provides direct information about the time and scale (frequency) location of the wheeze.

Wheeze denoising. The denoising performance of CWT-WED is achieved by applying *scale-dependent thresholding* [102]. This helps to reduce the high- and low-frequency noise components in the reconstructed signal, because the inverse CWT (ICWT) of (2.51) is performed over a range of scales, that is, $a^\circ < a < a^\bullet$, where a° and $a^\bullet$ denote the cutoff scales for the high and low frequencies, respectively.

Due to the localization of CWT both in time and scale (frequency), the reconstructed signal resembles the output of an adaptive filtering (AF), retaining the desired structural characteristics only at the true locations of the signal of interest $s(t)$.

Figure 3.1 depicts an example of the application of CWT-WED to LS recordings, presenting the wheeze identification part of CWT-WED. The top subfigure shows the recorder signal (normalized amplitude) along with the superimposition of the normalized airway flow signal. This contributes to the identification of the inspiratory (negative flow) and expiratory phase (positive flow) of the breathing cycle and the characterization of wheezes as inspiratory or expiratory, respectively. The middle subfigure depicts the corresponding spectrogram, estimated using a 256-point Hanning windowed signal with 50% overlapping and a discrete Fourier transform length of 2048 points. The bottom subfigure illustrates the scalogram (see (3.3)) from the estimated CWT of the analyzed LS signals. Figure 3.1 shows that the scalogram (bottom), unlike the spectrogram (middle), provides a timescale (frequency) representation of the wheezes that eliminates most of the

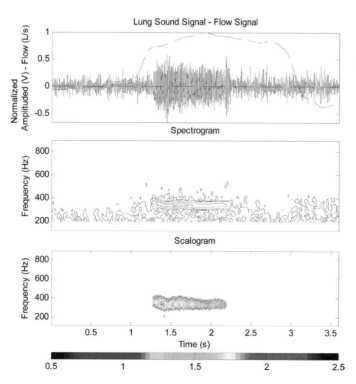

FIGURE 3.1: Wheeze detection functionality of CWT-WED [158]. (top) Recorded LS; (middle) corresponding STFT; (bottom) corresponding CWT.

undesired information related to background noise. As a result, the time–frequency characteristics of wheezing episodes are clearly revealed and identified.

Figure 3.2a depicts the CWT-WED wheeze denoising functionality. In particular, the top, middle, and bottom subfigures present the recorded lung sound (zoomed version of the recording in Figure 3.1), estimated wheeze, and breathing sound (background noise), respectively. In this figure, the adaptive character of CWT-WED in isolating the appropriate portion of the signal that corresponds to the wheeze location in time is evident. This is further justified by the corresponding power spectrum (PS) of the three signals included in Figure 3.2a, as shown in Figure 3.2b. From the latter, the dominant frequency of the wheeze that is evident in the PS of the LS recording (solid line) totally coincides with the main content of the PS of the estimated wheeze (dash-dotted line). Clearly, the PS of the remaining breathing sound (dashed line) follows the morphology of the original signal in the area outside the location of the dominant frequency peak of the wheeze. This proves the efficiency of CWT-WED not only in locating but also in extracting the signal of interest from the background noise.

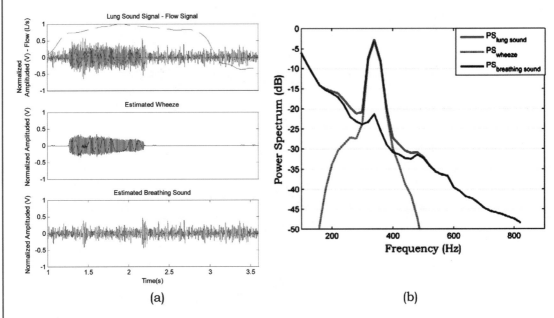

(a)   (b)

FIGURE 3.2: Wheeze denoising functionality of CWT-WED [158]. (a) The top part denotes the recorded LS; the middle part illustrates the corresponding estimated wheeze episode; the bottom part shows the corresponding breathing sound (background noise), (b) Estimated PS corresponding to the three signals in panel (a); wheeze denoising with CWT-WED is clearly reflected in the frequency domain.

3.2.2   WTST-NST Filter

Automated DAS detection and isolation from VS is a difficult task, due to DAS characteristics (Section 1.2). To address this problem, the nonstationarity of DAS must be taken into account. Consequently, the use of high-pass filtering (HPF) fails to separate the nonstationary sounds, destroying the waveforms. Furthermore, a level slicer cannot overcome the small amplitude of fine crackles (FC). Application of time-expanded waveform analysis in crackle time domain analysis [43, 159] results in separation; it is, however, time consuming and has large interobserver variability. Nonlinear processing [44] yields more accurate results, but requires empirical definition of the set of parameters of its stationary–nonstationary (ST–NST) filter.

As discussed in Section 2.4, WT provides a new perspective in the analysis of LS, because it can decompose them into multiscale details, describing their power at each scale and position [108]. By applying a threshold-based criterion at each scale, a filtering scheme which weights WT coefficients according to signal structure can be composed. Separation of signal from "noise" can be achieved through an iterative reconstruction–decomposition process, based on the derived weighted WT coefficients at each iteration.

The aforementioned concept was introduced in [46], where the implementation of a WTST–NST filter for the separation of DAS (nonstationary waves) from vesicular ones (stationary waves) is described. WTST-NST is a wavelet domain filtering technique, based on the fact that explosive peaks (DAS) have large components over many wavelet scales, whereas "noisy" background (VS) dies out swiftly with increasing scale. This fact allows characterization of the wavelet transform coefficients with respect to their amplitude; the most significant coefficients at each scale, with amplitude exceeding some threshold, correspond to DAS, whereas the rest correspond to VS. Consequently, a wavelet domain separation of wavelet transform coefficients corresponding to DAS and VS, respectively, can offer a time domain separation of DAS from VS using an iterative multiresolution decomposition–multiresolution reconstruction (MRD–MRR) scheme (Section 2.4). Although VS are characterized as "noisy" background, it is better to view them as an N-sample signal being incoherent relative to a basis of waveforms—that is, it does not correlate well with the waveforms of the basis, that is, its entropy is of the same order of magnitude as $\log N - \varepsilon$, with small ε [160]. From this notion, the separation of DAS from VS becomes a matter of coherent structure extraction from the breath sounds.

Figure 3.3 illustrates an example of the application of the WTST-NST filter to recorded LS that contain coarse crackles (CC). In Figure 3.3, the top subplot illustrates the original recorded LS sounds; the nonstationary character of CC is clearly evident. The result of the application of the WTST-NST filter to the recorded LS is depicted in the middle subplot of Figure 3.3, whereas the estimated VS signal (serving as background noise) is shown in the bottom subplot of Figure 3.3.

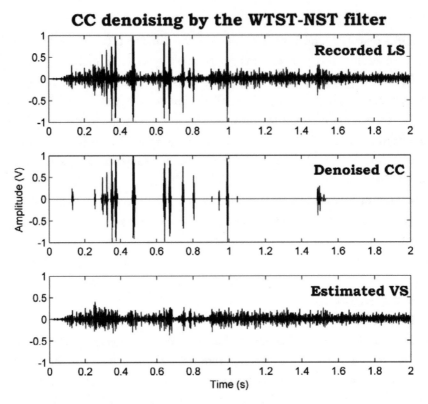

FIGURE 3.3: An example of the application of the WTST-NST filter [46] to recorded LS that contain CC. (top) Original recorded LS; (middle) denoised CC; (bottom) estimated VS (background noise-like). Results shown in this figure illustrate the efficient performance of the WTST-NST filter to identify and extract DAS from VS.

When comparing the output of the WTST-NST filter with the original signal, the ability of the WTST-NST filter to efficiently identify and extract DAS from VS is realized. Extended implementation details and further experimental results can be found in [46].

3.3   KURTOSIS-BASED EXTRACTOR

Higher-order statistics (HOS)-based analysis of LS has proved to be highly efficient as a new representation domain for the extraction of LS characteristics (Section 2.2). In this vein, HOS has also been used as basis for the construction of a denoising tool, especially for DAS. This was introduced in [56], where an iterative kurtosis-based detector (IKD) was formed.

The IKD scheme is based on the estimation of the kurtosis [see (2.6)] within a sliding window along the LS recordings. Because kurtosis is (theoretically) zero for Gaussian signals [86], such as VS (DAS background noise), significant deviations from this value can be attributed to the presence of non-Gaussian signals, such as DAS. This deviation from zero value can be used in forming a criterion for identifying the presence of nonstationary transient signals (i.e., DAS). The IKD segments the signal into two regions: one that consists of DAS plus VS, and one that consists only of VS. Unlike IKD, a detector based on variance changes fails in cases where Gaussian noise is present.

In general, for an N-sample sequence $\mathbf{x} = \{x(k): k = 1, 2, \ldots, N\}$, such as real observations, the estimate of the normalized kurtosis [see (2.6)] (the superscript x is omitted for simplicity) is calculated as [161]

$$\widehat{\gamma}_4 = \frac{\sum\limits_{k=1}^{N} [x(k) - \widehat{m}]^4}{(N-1)\widehat{\sigma}^4} - 3, \tag{3.4}$$

where $\widehat{m}$ and $\widehat{\sigma}$ denote the estimates of the mean and standard deviation of $\mathbf{x}$, respectively.

Kurtosis is a measure of the heaviness of the tails in the probability density function of $\mathbf{x}$ [86], which changes due to nonstationarity. In particular, outliers of abrupt changes in $\mathbf{x}$ sequence affect the tails of the probability density function. Hence, estimation of kurtosis via (3.4) could be used to establish an effective statistical test to identify abrupt changes in signals, such as those that characterize the existence of DAS in LS recordings.

The IKD scheme is implemented in an iterative form. In particular, along the given sound recording, $\mathbf{x}$, an M-sample sliding window ($M \ll N$) is defined, with one sample difference between successive windows. Over each sound segment obtained from the sliding window, the kurtosis $\widehat{\gamma}_4$ is estimated using (3.4). Afterwards, the derived time series $\widehat{\gamma}_4 = \{\widehat{\gamma}_{4,j} : j = 1, 2, \ldots, N - M + 1\}$ is demeaned and its absolute value is obtained, defining the time series $\mathbf{K} = \{K_j\}$. The time series $\mathbf{K}$ is then compared to a threshold, which is proportional to the standard deviation of $\mathbf{K}$ normalized by the standard deviation of the estimated VS that initially coincides with the sound recording $\mathbf{x}$. This normalization results in the adjustment of the threshold according to the mean power of the estimated VS and it tunes the threshold to efficient value as far as the detectability of the algorithm is concerned. The parts of the signal whose corresponding K_j value exceeds the threshold are assigned as signal DAS; the rest of the signal $\mathbf{x}$ is considered as the estimate of the VS.

However, by this approach, it is possible to have significant DAS portions that are not recognized as signal of interest and have been left in the VS part. Thus, the aforementioned procedure is iteratively applied as an effort to extract new DAS portions from the current VS signal. The iterative procedure is terminated when the mean power of the current DAS portion that is added to the previously estimated ones is smaller than a preselected power level $\varepsilon(\varepsilon \ll 1)$. The final DAS signal is the sum of all DAS portions estimated at all previous iterations, whereas the VS estimated at the last iteration qualifies as the final VS signal [56].

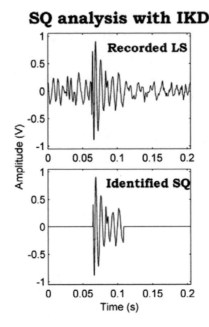

FIGURE 3.4: An example of the application of IKD [56] to recorded LS that contain SQ. (top) Original recorded LS; (bottom) detected SQ. Results shown in this figure illustrate the efficient performance of the IKD filter to detect DAS in LS recordings.

Figure 3.4 illustrates an example of the implementation of IKD to LS recordings that includes a squawk. As can be seen from this figure, IKD keeps the part of the signal that includes the DAS only, and outputs zeroes at places where only VS signal exists. Nevertheless, as already noted, the VS signal that coexists in the DAS portion is also extracted in the IKD output, meaning that, unlike the WTST-NST filter presented in Section 3.2, IKD efficiently allocates the portions of the LS recordings that include DAS and extracts them from the original signal, yet without separating DAS from VS in the extracted segments. However, as Figure 3.4 shows, IKD preserves the characteristics of squawks (SQ) in its output, because the two parts of SQ, that is, initial FC followed by a short wheeze, are clearly identified.

For more details about the performance of the IKD scheme, the reader is referred to [56].

3.4   FRACTAL DIMENSION-BASED DETECTOR

Apart from the fractal analysis of LS presented in Section 2.8, a fractal dimension (FD)-based detector is presented here, and applied to DAS detection within the LS recordings. In particular, the FD-based detector (FDD) introduced in [53] assesses the complexity of sound recordings in the time domain in order to efficiently detect the time location and duration of the nonstationary DAS in LS data.

The FDD scheme is based on the FD of the LS recordings. As discussed in Section 2.8, FD could be used as a measure of signal complexity in the time domain. This complexity could vary with different pathophysiological conditions. Consequently, by means of FD the complexity variations are linked with the changes over time of sound recordings, providing a fast computational tool that tracks the nonstationarities, that is, the location of the explosive DAS.

The FDD scheme initially employs an *M*-sample sliding window, which is shifted along the *N*-sample section of the LS recordings, with a 99% percentage of overlap, obtaining point-to-point values of the estimated FD. Over each sound segment obtained from the sliding window, the FD is computed using Katz's algorithm [see (2.82)], and the value of the estimated FD^K is assigned to the midpoint of the sliding window. When FD^K is estimated, an FD-peak peeling algorithm (FD-PPA) [53] is used to automatically identify the location and duration of the explosive DAS through the FD peaks. The FD-PPA iteratively "peels" the estimated FD^K signal, gradually gathering those parts that construct its peaks. In this way, even small-amplitude FD^K peaks are accurately identified. In the unlikely case with merged DAS peaks, which have not been successfully separated by the first pass of the FD-PPA, the latter is applied again locally; hence, with one or two passes of the FD-PPA, the time location and duration of all explosive DAS are accurately and automatically identified [53].

Figure 3.5 shows an example of the application of FDD to LS recordings that include FC. In this figure, the top subfigure illustrates the original LS recordings with FC superimposed on the VS

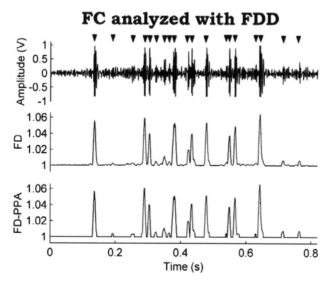

FIGURE 3.5: An example of the application of FDD [53] to recorded LS that contain FC (denoted by arrowheads). (top) Original recorded LS; (middle) estimated FD^K; (bottom) output of the FD-PPA corresponding to valid FC peaks.

sound. The middle subfigure denotes the estimated FD^K, whereas the bottom subfigure shows the output of the FD-PPA applied to the estimated FD^K, keeping only the FD^K peaks that correspond to valid FC ones. As shown in Figure 3.5, variation in the amplitude of FC peaks does not affect the performance of the FDD scheme, because it efficiently detects all FC—despite their high, medium, or low amplitude—compared to the VS amplitude level.

The FDD technique has the potential advantage of data volume reduction. This could be realized by posing criteria (e.g., threshold values) to the estimated FD^K values, which could help an expert system in deciding whether to store or discard the incoming signal included in the analysis window. This is of great importance when LS are continuously recorded for long-term analysis.

More details about the FDD scheme could be found in [53].

3.5 WAVELET-FRACTAL DIMENSION-BASED DENOISING

In Sections 3.2 and 3.4, denoising techniques based on WT and FD, respectively, were presented. WT-FD filter, a combinatory approach that merges the advantages of WT and FD analysis, was introduced [54, 55] in order to achieve superior performance in the enhancement of DAS. In particular, the efficiency of the FDD to accurately locate DAS in LS recordings (Section 3.4) is transferred to the WT domain as a means that values the WT coefficients according to their significance in the signal structure. Consequently, an FD-based thresholding procedure is formed, resulting, through an iterative reconstruction–decomposition process, in efficient separation of the desired signal (i.e., DAS) from the undesired VS (background noise).

The main thrust of the WT-FD filter is the more enhanced selection of WT coefficients that correspond to DAS during the MRD–MRR procedure (Section 2.4), compared to that of the WTST-NST filter (Section 3.2). The WT-FD filter overcomes the requirement of the WTST-NST filter for empirical setting of a multiplicative parameter in the definition of its threshold [46], providing a different perspective in the categorization of the WT coefficients by using FD analysis to construct an efficient method of thresholding. As explained in Section 3.4, FDD enables the detection of DAS, but not their extraction from background noise. In the WT-FD filter, the detection ability of the FDD scheme is transferred from the time domain to the WT domain. In this manner, both detection and extraction of DAS signals from background noise are achieved, by reconstructing the signal of interest (DAS) through the WT coefficients efficiently selected by the FDD.

Using FD-based thresholding, the WT-FD filter acts *only* at the true locations of the desired signal presence, keeping the rest of the input signal unchanged. Consequently, similar to the WTST-NST filter (Section 3.2), the WT-FD filter also performs as an *adaptive* noise removal tool for the analysis of LS.

Figure 3.6 illustrates an example of the application of the WT-FD filter to LS recordings containing CC. As shown in this figure, the efficiency of the FDD scheme to identify the useful

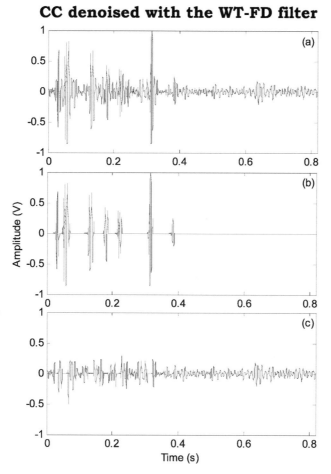

CC denoised with the WT-FD filter

FIGURE 3.6: An example of the application of the WT-FD filter [54, 55] to recorded LS that contain CC. (a) Original recorded LS; (b) denoised CC; (c) estimated VS (background noise-like).

WT coefficients that correspond to valid DAS peaks has resulted in accurate extraction of CC (Figure 3.6b) from the original LS recordings (Figure 3.6a). The remaining signal (Figure 3.6c) qualifies as VS and/or background noise. A thorough description of the structure and performance of the WT-FD filter can be found in [54, 55].

3.6 EMPIRICAL MODE DECOMPOSITION-FRACTAL DIMENSION-BASED DENOISING

An extension of the combinatory approach used in the WT-FD filter to the field of empirical mode decomposition (EMD; Section 2.7) is introduced in [57]. Instead of WT, the EMD is used to decompose the sound signal into components with well-defined instantaneous frequency. In this manner, the oscillatory characteristics of DAS are reflected to the intrinsic mode function (IMF) analysis domain. Consequently, by applying FD analysis in the latter, the important (high FD value) and unimportant (low FD value) portions of IMFs can be identified, corresponding to DAS and background noise, respectively. Because none of the signal is lost in EMD, the sum of the selected portions of IMFs per category gives back the denoised DAS and the background noise, accordingly.

The aforementioned approach forms a new denoising tool, the EMD-FD filter, which adaptively captures the nonstationarity of ELS and successfully extracts them from the background noise. According to the desired accuracy in its output, the EMD-FD filter is formed in two ways: in a noniterative and an iterative way.

In the noniterative realization of the EM-FD filter, an EMD process is applied to the M-sample LS signal $x(k)$, $k = 1, 2, \ldots, M$, decomposing it into N IMFs, that is, $c_i(k)$, $k = 1, 2, \ldots, N$, and the residue $r_N(k)$, $k = 1, 2, \ldots, M$, following the procedure described in EMD analysis (Section 2.7). From the N estimated IMFs, the first L ones are selected according to an energy criterion [57]. The FDD scheme (Section 3.4) is then applied to the L selected IMFs, and for each one, the valid peaks of the corresponding FD sequences, $FD_j^K(k)$, $j = 1, 2, \ldots, L$, $k = 1, 2, \ldots, M$, are estimated, that is, $FDPP_j^K(k)$, $j = 1, 2, \ldots, L$, $k = 1, 2, \ldots, M$, picked by using the FD-PPA scheme within the FDD (Section 3.4). After the generation of the $FDPP_j^K(k)$ sequence, two binary thresholds are constructed, signal binary threshold $SBTH_j^K(k)$ and noise binary threshold $NBTH_j^K(k)$, defined as follows:

$$SBTH_j(k) = \begin{cases} 1 & FDPP_j(k) \neq 1 \\ 0 & FDPP_j(k) = 1 \end{cases}, \qquad (3.5)$$

$$NBTH_j(k) = \left[1 - SBTH_j(k) \right], j = 1, 2, \ldots, L, \; k = 1, 2, \ldots, M. \qquad (3.6)$$

When the binary thresholds of (3.5) and (3.6) are constructed, they are both multiplied with the corresponding IMFs. In this way, by means of $SBTH_j(k)$, the portions of IMFs that are related to the desired signal (DAS), $c_j^{DAS}(k)$, $j = 1, 2, \ldots, L$, $k = 1, 2, \ldots, M$, are kept, whereas using the $NBTH_j(k)$, those related to the background noise, $c_j^{BN}(k)$, $j = 1, 2, \ldots, L$, $k = 1, 2, \ldots, M$, are obtained. By combining all coherent parts from L IMFs, the estimated denoised DAS signal is produced as

$$x^{DAS}(k) = \sum_{j=1}^{L} c_j^{DAS}(k), \; k = 1, 2, \ldots, M, \qquad (3.7)$$

whereas the noncoherent part (background noise) is estimated as the summation of the remains, that is,

$$x^{\mathrm{BN}}(k) = \sum_{j=1}^{L} c_j^{\mathrm{BN}}(k) + \sum_{i=L+1}^{N} c_i(k) + r_N(k), \; k = 1, 2, \dots, M.$$

$$(3.8)$$

The noniterative structure presented above can be further expanded to an iterative one, when the $x^{\mathrm{DAS}}(k)$ output [see (3.7)] is not of the desired accuracy. This means that some portions of the DAS signal still exist in $x^{\mathrm{BN}}(k)$ [see (3.8)]. This is probably due to the amplitude-overshadowing of some DAS peaks over some others, yet still valid. In other words, some high-amplitude FD peaks that correspond to high-amplitude DAS peaks have led to an underestimation of the FD peaks that correspond to the lower-amplitude DAS peaks, resulting in a misidentification by the FDD scheme. To circumvent this problem and to provide refinement alternatives within the performance of the EMD-FD filter, an iterative structure of it was formed [57]. In this EMD-FD realization, the $x^{\mathrm{BN}}(k)$ derived by (3.8) is used iteratively as an input signal to the EMD-FD analysis. At each iteration, an energy-based stopping criterion is applied [57], and its convergence to e, where e is a small positive number ($0 \ll \varepsilon \ll 1$) that corresponds to the desired accuracy in the refinement procedure, is tested. When convergence is achieved and the stopping criterion is met after the Cth iteration, the final output of the EMD-FD filter—denoised DAS, $x_{it}^{\mathrm{DAS}}(k)$, and background noise, $x_{it}^{\mathrm{BN}}(k)$—are given as

$$x_{it}^{\mathrm{DAS}}(k) = \sum_{l=1}^{C} x_l^{\mathrm{DAS}}(k) \quad \text{and} \quad x_{it}^{\mathrm{BN}}(k) = x_C^{\mathrm{BN}}(k), \; k = 1, 2, \dots, M,$$

$$(3.9)$$

where $x_l^{\mathrm{DAS}}(k)$ corresponds to the estimated denoised DAS signal [similar to (3.7)] at the lth iteration.

From the procedure above, it is clear that the denoised DAS signal is formed bit by bit, adding all the estimated DAS parts derived at each iteration. In addition, the background noise estimated at the Cth iteration qualifies as the final version of the background noise, because further analysis contributes nothing more to the refinement procedure.

Figure 3.7 depicts an example of the application of the EMD-FD filter to LS recordings that include two SQ (denoted with arrowheads). As can be deduced from this figure, denoised SQ (Figure 3.7b) are clearly separated from the original LS recordings (Figure 3.7a), whereas the remaining signal (Figure 3.7c) resembles the characteristics of the background noise (e.g., compare it with portions of LS recordings in Figure 3.7a outside the regions of the SQ existence). Moreover, in the denoised version of SQ (Figure 3.7b), it is clear that the EMD-FD filter sustains the morphology of the SQ. This is important, because SQ include both an FC and a short wheeze [12]; hence, not only their initial part corresponding to FC but the part with the fluctuation corresponding to the short wheeze should be identified and separated from the background noise. This is clearly achieved

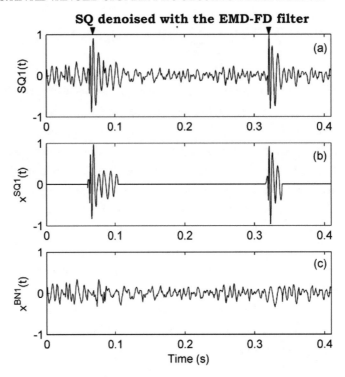

FIGURE 3.7: An example of the application of EMD-FD filter [57] to recorded LS that contain SQ. (a) Original recorded LS; (b) denoised SQ ; (c) estimated background noise.

by the EMD-FD filter, as it can be deduced from Figure 3.7b, justifying the adaptive performance of the EMD-FD filter.

For an analytical description of the EMD-FD filter the reader should refer to [57].

3.7   HEART SOUND CANCELLATION

Heart sounds produce an incessant noise during LS recordings that causes an intrusive, quasi-periodic interference that influences the clinical auscultative interpretation of LS. The introduction of pseudo-periodicities, the masking of the relevant signal and the modification of the energy distribution in the spectrum of LS, due to heart sounds [30], require an effective reduction of heart sounds from the contaminated LS signal, to yield a successful LS classification.

HPF and AF are the two main approaches for heart noise reduction. Although HPF (with a cutoff frequency varying from 50 to 150 Hz) is effective in heart sound reduction [17, 162], it degrades the respectively overlapped frequency region of breath sounds and fails to track the changing signal characteristics. The AF technique overcomes these limitations, because it is based on a gradual reduction of the mean square error between the primary input signal (contaminated LS) and a recorded or artificially produced reference signal, highly correlated to the noise component of the input signal (heart sounds) [30, 34, 163]. This approach was initially proposed by Iyer et al. [30], who used AF with a noise reference signal derived from a modified electrocardiogram signal. A modification to this method, proposed by Kompis and Russi [34], combined the advantages of AF with the convenience of using only a single microphone input. Unfortunately, this approach has resulted in a moderate heart sound reduction by 24%–49%. Stemming from these two initial efforts, more efficient heart sound reduction techniques have been developed over the years, using advanced signal processing methodologies. These can be further categorized into those that analyze the entire LS record and reduce heart sounds effect, and those that remove heart sounds included in portions of the LS record and then estimate the signal in the gaps [25]. Several indicative examples of heart sound cancellers (HSCs) from both categories are described in the succeeding subsections.

3.7.1 HOS-Based HSC

HOS have been used in the construction of an HSC in order to adaptively reduce the heart sound noise from LS recordings. In particular, an adaptive noise canceller based on fourth-order statistics (ANC-FOS) was introduced in [42], as a means to form an efficient HSC.

ANC-FOS has two main parts; the first one analyzes the incoming breath signal $x(k)$ to generate the reference signal $z(k)$ for the adaptive filter, whereas the second one computes the FOS of $x(k)$ and $z(k)$, forms the adaptive filter, and yields the final output. To generate the reference signal, the real location of the heart sound in the incoming breath signal must be detected. To this end, in the first part of the ANC-FOS $x(k)$ is analyzed and searched for the true locations of heart sound noise, based on amplitude, distance, and noise-reduction percentage criteria [42]. Its output, $z(k)$, is a localized signal, with precise tracking of the first and second heartbeats, highly correlated with heart sounds, with no extra recording requirement as in Iyer et al.'s method [30]. The second part of ANC-FOS includes the adaptive section of the algorithm and starts with the computation of the fourth-order cumulants of $z(k)$, $c_4^{zzzz}(\tau_1,\tau_2,\tau_3)$, and the fourth-order cross-cumulants of $x(k)$ and $z(k)$, $c_4^{zzzz}(\tau_1,\tau_2,\tau_3)$ (Section 2.2), assuming that they exist and they are not identically zero. These FOS are then used to form the adaptive-filter update equation, given by

$$H_f(k+1) = H_f(k) - \mu(k)\nabla(k),$$
(3.10)

with

$$H_f = [h(0), h(1), \ldots, h(N-1)]^T,$$
(3.11)

and

$$\nabla \equiv 2 \left(c_4^{zzzzT} c_4^{zzzz} H_f - c_4^{zzzzT} c_4^{xzzz} \right), \tag{3.12}$$

where $\boldsymbol{h}(i)$, $i=0, ..., N-1$, denotes the adaptive filter coefficients (N taps) and $m(k)$ corresponds to the step size [42]; bold letters indicate the use of vectors.

Given that the FOS of Gaussian processes are identically zero (Section 2.2), the update equation (3.10) of the adaptive filter consists only of the fourth-order cumulants of the incoming and

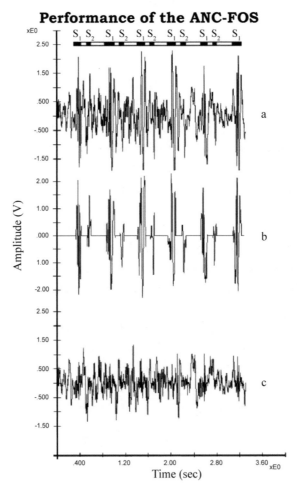

FIGURE 3.8: An example of the application of ANC-FOS [42] to recorded LS, severely contaminated with heart sound noise. (a) Original recorded LS; (b) created noise-reference signal $z(k)$; (c) denoised LS. S_1 and S_2 indicate the first and second heart sounds, respectively.

reference signals and is not affected by Gaussian uncorrelated noise. In addition, ANC-FOS has a localized effect on the input signal at locations pointed out from the reference signal, unlike HPF that deteriorates the whole signal due to elimination of low frequencies.

Figure 3.8 illustrates an example of heart sound cancellation in LS recordings during normal breathing when using the ANC-FOS scheme. When comparing Figure 3.8a and b, it is realized that the reference signal $z(k)$ is highly correlated to the heart sounds existing in the initial LS recording (Figure 3.8a) and exists at the time instances of the heart sounds only. Clearly, in the denoised output of the ANC-FOS (Figure 3.8c), the heart sound interference has been removed, yet without destroying the coexisting LS.

Further details about the ANC-FOS can be found in [42].

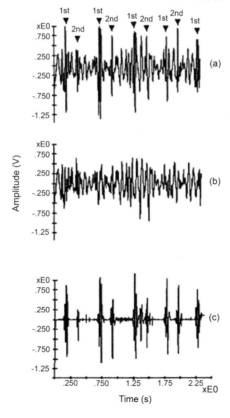

FIGURE 3.9: An example of the application of WTST-NST as HSC [28] to recorded LS, severely contaminated with heart sound noise. (a) Original recorded LS; (b) denoised LS; (c) estimated heart sound interference. Arrowheads denote the first and second heart sounds.

3.7.2   WT-Based HSC

A WT-based adaptive denoising of LS contaminated by heart sound noise is presented in [28]. In fact, this approach refers to the application of the WTST-NST filter (Section 3.2) to the problem of heart sound cancellation. The adaptive separation of signal from "noise," without requiring any reference signal, can be achieved through the iterative reconstruction–decomposition process embedded in the WTST-NST filter, based on the derived weighted WT coefficients at each iteration.

In the realization of the WTST-NST filter as HSC, the DAS, which were originally the signal of interest, are replaced by the heart sound peaks, whereas the originally estimated background noise qualifies here as the desired signal of interest, that is, the denoised LS.

Figure 3.9 presents an example of the performance of the WTST-NST filter as HSC, when applied to severely contaminated LS. As Figure 3.9c shows, when compared to Figure 3.9a, the heart peak locations in input and in heart sound WTST-NST output are synchronized, without using any reference signal. Furthermore, when comparing Figure 3.9a and b, it can be seen that the signal included between the heart sounds remains unchanged. Consequently, the WTST–NST filter when used as an HSC has a localized effect on the input signal, at true locations of heart sound presence without requiring any reference signal.

More details about the WTST-NST filter acting as a HSC can be found in [28].

3.7.3   Recursive Least Squares-Based HSC

Recursive least squares (RLS) have been proposed for linear AF of heart sound noise by Gnitecki et al. [35]. They used a bandpass filtered version of the recorded LS signal to initially detect the presence of heart sound peaks (following Moiussavi et al. [164]) and then they replaced all the components of the bandpass filtered original signal by zero, except for the heart sound that included parts. The resulting impulse-like signal was used as the noise-reference signal for the RLS-based HSC. The latter tries to find the component in the main signal that matches with the noise reference and performs filter coefficient optimization using RLS. The adapted-to-noise output is then subtracted from the original LS recordings, resulting in the denoised version of LS; more details can be found in [35].

3.7.4   Time–Frequency-Based HSC

The use of short-time Fourier transform (STFT) as a means to construct an HSC was presented by Pourazad et al. [37]. In fact, the main concept was to use a time-varying filtering based on STFT to remove heart sound segments from LS records and then by using simple interpolation to fill the gaps in the time–frequency plane. The denoised LS are then produced via the inverse STFT; more details can be found in [37].

An extension to the above-mentioned approach was introduced in [39, 40], where heart sound segments are detected by multiscale product and removed from the WT coefficients at every scale. Removal of heart sound segments is followed by an estimation of the missing data via linear prediction using autoregressive (AR) or moving average (MA) models.

As reported in [39, 40], the choice of AR or MA modeling depends on the location of the heart sound segment within the breathing cycle; for heart sound segments close to the onset of inspiration or expiration, MA modeling is preferred, because most of the information about the samples in the gap is in the next lung sound segment; otherwise, AR modeling is preferred. For a more complete description, the reader is referred to [39, 40].

3.7.5   Recurrent Time Statistics and Nonlinear Prediction-Based HSC

Another HSC approach, which is based on a recurrence time statistic combined with nonlinear prediction, was introduced by Ahlstrom et al. [41]. They explored the sensitivity of a recurrence time statistic to changes in a reconstructed state space (multivariate vector space). Signal segments that are found to contain heart sounds are removed, and the arising missing parts are replaced with predicted LS by using a nonlinear forward and backward prediction scheme, dividing the missing segment in two equal halves.

The prediction operates in the reconstructed state space and uses an iterated integrated nearest trajectory algorithm. Similar trajectories in state space share the same waveform characteristics in time domain, and a way of predicting missing data is thus to mimic the evolution of neighboring trajectories. Six LS segments surrounding the heart sound segment were used to reconstruct the state space, and five nearest neighbors were used in the prediction.

The predicted reconstructed data are then brought back into the time domain as the first coordinate of the embedding vector. The authors report auditory high-quality results, with successful removal of heart sounds from the LS recordings, yet with some shifting effect to higher frequencies in the denoised LS. Analytical description of the performance of the recurrent time statistics and nonlinear prediction-based HSC can be found in [41].

3.8   OTHER DENOISING APPROACHES

Apart from the aforementioned denoising/HSC techniques, a few other techniques have been reported in the literature; they are briefly described here for the sake of completeness.

3.8.1   Fuzzy Logic-Based Denoising

Throughout the years, a whole family of LS denoising techniques has been based on the use of fuzzy logic, aiming at real-time analysis of LS. In particular, the fuzzy rule-based stationary–

nonstationary (FST-NST) filter has been presented by Tolias et al. [47], in an attempt to construct an enhanced real-time separation scheme of DAS (nonstationary waves) from VS (stationary waves). The FST-NST filter consists of two adaptive neurofuzzy inference systems [165] that operate in parallel, for the estimation of the stationary and the nonstationary waves for crackles. An alternative approach for real-time separation of DAS has been suggested by Mastorocostas et al. [48] with the orthogonal least squares method fuzzy filter (OLS-FF). This filter also consists of two fuzzy inference systems that operate in parallel. A complete model-building process is also proposed, based on the orthogonal least squares method. The OLS-FF performs separation of crackles and squawks.

The above-mentioned neurofuzzy models are static models, which attempt to capture the plant dynamics by feeding the networks with delayed values of the input signal. However, for identification of temporal processes, dynamic models have been used, exhibiting some type of memory. The recurrent models are able to learn the system dynamics without assuming much knowledge about the structure of the system under consideration. In this vein, a filter that was based on the dynamic fuzzy neural network (DFNN) [166] was constructed for the separation of all types of DAS [167]. DFNN is a recurrent model and is capable of effectively identifying the dynamics of a process. A recurrent neural filter has been reported in [168], based on the block-diagonal recurrent neural network (BDRNN) [169], which is a simplified form of the fully recurrent network, with no interlinks among neurons in the hidden layer. As shown in [168], due to its internal dynamics, the BDRNN filter is capable of performing efficient real-time separation of DAS from VS.

3.8.2   Independent Component Analysis-Based HSC

The recovery of independent source signals, when a linear mixture of signals is only available with unknown sources and mixing processes, is often addressed by using independent component analysis (ICA). The latter finds a linear coordinate system where the recovered signals are statistically independent [170]. For cases where signals are not linearly mixed but convolved, application of ICA to the spectrograms of the mixed signals (at least two versions are required) provides a satisfactory solution, assuming that the source signals are originally statistically independent and at least one of them is Gaussian. An ICA-based HSC was proposed in [36], where two recordings of LS contaminated with heart sounds were assumed as the mixed signals. By combining the proper independent components from each frequency derived after the ICA, the spectrograms of the separated signals were produced. Using inverse STFT (ISTFT), the separated signals were reconstructed back into the time domain. As Porazad et al. report [36], the proposed HSC method reduces the effect of heart sound noise, yet a weakened heart sound could still be heard in the resultant signal along with a noticeable acoustic change in the LS portions free of heart sounds.

3.8.3 Variance Fractal Dimension-Based Heart Sound Localization

A variance fractal dimension (VFD)-based (Section 2.8) heart sound localization technique was proposed in [171]. The VFD trajectory was created by calculating the VFD of the recorded LS in a moving window of 100 ms with an increment size of 25 ms. The VFD trajectory was then compared with an adaptive threshold (tuned for every subject); the segments above the threshold were considered as those that contain heart sounds. This approach exhibits increased computational burden and its accuracy depends on the window size of VFD calculation [171].

3.8.4 Entropy-Based Heart Sound Localization

A Shannon entropy-based heart sound localization technique was proposed in [27]. This method uses the statistical information that the entropy provides to differentiate between LS segments with and without the presence of heart sounds. In fact, due to changes in the probability density function, the entropy of the LS segments with heart sounds is greater than that of the segments without heart sounds. Mean plus standard deviation of the estimated entropy was used a threshold for the detection of heart sound locations; the entropy of the latter takes values that exceed the threshold. Yadollahi and Moussavi state [27], this approach exhibits high heart sound detectability, high resolution in boundary locations of the heart sound, and robustness to variations between subjects.

For an analytical comparison among ICA-based HSC, and VFD- and entropy-based heart sound localization methods, the reader is referred to the work of Moussavi [25, Section 5.3].

CHAPTER 4

Reflective Implications

4.1   FROM AN ENGINEER'S VIEWPOINT

The new analysis domains and the denoising methodologies presented in the previous two chapters, respectively, show the potentiality of the advanced signal processing techniques to offer a wide range of new opportunities in lung sound (LS) analysis. Apparently, this evolution of capabilities is related to a series of parallel developments, such as technology growth, introduction of new theories, better understanding of the mechanisms of biological function, gradual integration of the biomedical engineering research findings in clinical practice. Therefore, the development of technologies related to medical care, health, and welfare are considered extremely crucial. In addition to expectations for the development of medical technologies arising from needs of the global economy, there are also increasing expectations for the creation of new medical technologies stemming from trends in development and practical application of the life sciences. As a result, developments in science and technology focusing on humans have become one of the mainstream trends.

Progress in biomedical engineering in the 20th century has greatly transformed the concept of health care itself, by dramatically transfiguring it from a form of classical medical care to that of leading edge medical care assisted by instrumentation and imaging. Furthermore, social demands to enhance and maintain a high quality of medical treatment, health, and welfare in this aging society are behind the anticipation for further developments in this field in the future. Development of biomedical instrumentation technology to measure the physical and chemical information and signals emitted by the living body in a minimally (or non-)invasive and continuous manner, was responsible for the establishment of the concept of patient monitoring within clinical medicine. It thus contributed greatly to the enhancement of the quality of medical care. Nevertheless, such perspective still lacks in the area of bioacoustics. Except in a few rare cases [23, 24, 172], continuous monitoring of LS (and bioacoustic signals, in general, that is, heart sounds, bowel sounds, joint sounds) is totally absent from clinical practice. Physicians use their temporal memory to store "acoustic impression" of their patients, which is quickly banished by a new one as the next patient comes in. This reality shows that although auscultation is one of the oldest diagnostic procedures still in practice, it has not followed the same rhythm of technology-based enhancement. The biomedical engineer who deals with bioacoustics finds her-/himself in an ambiguous situation. On one hand, (s)he finds difficulties in gathering systematic bioacoustic experimental data, due to the lack of everyday recordings in hospitals. On the other hand, the slow development curve of bioacoustics analysis itself provides an open field for further research and exploitation. The latter could be successfully realized through an interdisciplinary approach, that is, by combining new signal processing and modeling techniques with the interpretation of bioacoustic data by physicians. Results amassed from the interdisciplinary field of engineering and medicine are beneficial in the maintenance of life, in particular, human life, and improvement of the environment.

The previous chapters made it clear that the intrinsic characteristics of LS, such as embedded periodicity (musical LS), nonlinearity, nonstationarity, and non-Gaussianity, require appropriate techniques that efficiently deal with LS properties in a straightforward manner. The combination of processing schemes with appropriate modeling could result in a more effective performance. This could be further enhanced if knowledge from biomechanical dynamic models of the cardiovascular system is transferred to the accompanying filed of sound modeling. Integration between signal processing and such modeling would be capable of directly attributing pathophysiological meaning to parameters obtained from signal processing algorithms.

4.2 FROM A PHYSICIAN'S VIEWPOINT

Sir A. C. Doyle (the writer of *Adventures of Sherlock Holmes* stories) in his *Scandal in Bohemia* offers a glimpse of the medical customs of the late 1800s; he wrote:

> As to your practice, if a gentleman walks into my rooms, smelling of iodoform, with a black mark of nitrate of silver upon his right forefinger, and a bulge on the side of his top-hat to show where he has secreted his stethoscope, I must be dull indeed if I do not pronounce him to be an active member of the medical profession.

Note that men's clothing of the day did not have pockets large enough for the Laennec stethoscope; therefore, it was unscrewed and carried in the physician's top hat [173]. This quotation reveals two things: the role of the stethoscope as a physician's hallmark and the necessity of its portability.

Modern-day physicians still sustain such relationship with the stethoscope. The only changes are the developed form of the stethoscope and its placement around physician's neck or in the pocket of their white robe; combined with the latter, the stethoscope is still the doctor's badge. From this viewpoint, efforts toward an intelligent auscultation system seem like intrusions that could well be viewed as altering the physician's image, devaluating his/her expertise and undermining his/her intellectual contribution to understanding and treating diseases. Actually, the locus of control of medical decision making is a key factor in examining the effects of technology-based assistive tools on a physician's behavior. This suggests that increased physician reliance on computerized auscultation, which may shift control of decision making away from the individual physician, may further depersonalize the doctor–patient relationship. Adding to this, computer-aided decisions are frequently based on rules and criteria embedded in the programs unknown to physicians. Moreover, these decision rules become relatively immune to change. Furthermore, computerized decision support systems are based almost entirely on quantifiable data. Nonquantifiable information that cannot easily be processed by the computer is largely ignored. Intelligent computerized auscultation, however, could offer high-quality auscultative data and indicators that carry important diagnostic information, helping the physician to come up with a more reliable decision about the patient's pathology.

Toward an intelligent auscultation system, considerations such as sensor (or multiple sensors) type, graphical display, built-in memory, wireless data transmission, and adaptive noise reduction should be taken into account; most importantly, all these should be fitted into a small, lightweight, portable, handheld device (a Pocket PC-like). Through synchronization, handheld data can be backed up and managed on a server-like PC (e.g., running database/data mining applications), further increasing the ways in which information can be gathered and distributed. Having high-quality auscultative information readily available at the point of care can be extremely useful in view of the growing amount of medical information available, the increased expectations to follow guideline, and formulary restrictions, and the time limitations placed on physicians. Handheld-type stethoscopes can improve accuracy, help avoid medical errors, and achieve better patient outcomes. To attain higher interfacing of handheld-type stethoscopes with hospital and clinic systems both directly and wirelessly, measures will have to be instituted to assure adequate security and compliance with new, more stringent regulations about medical record privacy and access.

Clinical validation, real-time implementation, generalization, noise robustness, and appropriate training are some of the key issues that need to be considered for a realistic scenario of intelligent auscultation systems. Although the signal processing methods described in this book do not account for *all* these issues, they create a bed-set for more pragmatic exploitation of lung sounds and contribute toward the enhancement of their diagnostic value. The ample space for new discoveries and optimization is the driving force that keeps bioacoustic research alive; yet, physicians' support and collaboration are the catalyst for the success of such endeavor.

References

[1] V.S. McKusick, *Cardiovascular Sound in Health and Disease*, Williams & Wilkins, Baltimore, MD, p. 3 (1958).

[2] J. Rapoport, Laennec and the discovery of auscultation, *Isr. J. Med.*, **22**, 597–601 (1986).

[3] M. Foucault, *The Birth of the Clinic: An Archaeology of Medical Perception* (A. M. Sheridan Smith, Trans.), Pantheon Books, New York, NY (1973).

[4] J. Sterne, Mediate auscultation, the stethoscope, and the 'autopsy of the living': medicine's acoustic culture, *J. Med. Humanit.*, **22**(2), 115–136 (2001).

[5] A.B. Davis, *Medicine and its Technology: An Introduction to the History of Medical Instrumentation*, Greenwood Press, Westport, CT, pp. 88–89 (1981).

[6] J. Crary, *Techniques of the Observer: On Vision and Modernity in the Nineteenth Century*, MIT Press, Cambridge, MA, pp. 89–90 (1990).

[7] V.S. McKusick, *Cardiovascular Sound in Health and Disease*, Williams & Wilkins, Baltimore, MD, p. 13 (1958).

[8] S.S. Kraman, Vesicular (normal) lung sounds: how are they made, where do they come from and what do they mean?, *Semin. Respir. Med.*, **6**, 183–191 (1985). doi:10.1055/s-2007-1011496

[9] R.E. Thacker and S.S. Kraman, The prevalence of auscultatory crackles in subjects without lung disease, *Chest*, **81**(6), 672–674 (1982). doi:10.1378/chest.81.6.672

[10] P. Workum, S.K. Holford, E.A. Delbono, and R.L.H. Murphy, The prevalence and character of crackles (rales) in young women without significant lung disease, *Am. Rev. Respir. Dis.*, **126**(5), 921–923 (1982).

[11] A.J. Robertson, Rales, ronchi, and Laennec, *Lancet*, **1**, 417–423 (1957). doi:10.1016/S0140-6736(57)92359-0

[12] S.S. Kraman, *Lung Sounds: An Introduction to the Interpretation of Auscultatory Findings*, Amer. College of Chest Physicians, Northbrook, IL, workbook (1983).

[13] R.L.H. Murphy, Discontinuous adventitious lung sounds, *Semin. Respir. Med.*, **6**, 210–219 (1985). doi:10.1055/s-2007-1011499

[14] D.W. Cugell, Lung sound nomenclature, *Am. Rev. Respir. Dis.*, **136**, 1016 (1987).

[15] J.E. Earis, K. Marsh, M.G. Rearson, and C.M. Ogilvie, The inspiratory squawk in extrinsic allergic alveolitis and other pulmonary fibroses, *Thorax*, **37**(12), 923–936 (1982).

[16] N. Gavriely and D.W. Cugell, *Breath Sounds Methodology*, CRC Press, Boca Raton, FL, p. 2 (1995).

[17] N. Gavriely, Y. Palti, and G. Alroy, Spectral characteristics of normal breath sounds, *J. Appl. Physiol.*, **50**, 307–314 (1981).

[18] T. Katila, P. Piirila, K. Kallio, E. Paajanen, T. Rosqvist, and A.R.A. Sovijarvi, Original waveform of lung sound crackles: a case study of the effect of high-pass filtration, *J. Appl. Physiol.*, **71**(6), 2173–2177 (1991).

[19] R.L.H. Murphy, S.K. Holford, and W.C. Knowler, Visual lung-sound characterization by time-expanded wave-form analysis, *N. Engl. J. Med.*, **296**, 968–971 (1978).

[20] L. Vannuccini, J.E. Earis, P. Helistö, B.M.G. Cheetham, M. Rossi, A.R.A. Sovijärvi, and J. Vanderschoot, Capturing and preprocessing of respiratory sounds, *Eur. Respir. Rev.*, **10**(77), 616–620 (2000).

[21] M.J. Mussell, The need for standards in recording and analysing respiratory sounds, *Med. Biol. Eng. Comput.*, **30**, 129–139 (1992). doi:10.1007/BF02446121

[22] L. J. Hadjileontiadis, BlueSteth: a Bluetooth-based stethoscope for healthcare and educational applications, *IEEE Computer Society, CSIDC 2002 Final Report*, pp. 1–30 (2002).

[23] R.L.H. Murphy, Localization of chest sounds with 3D display and lung sound mapping, U.S. Patent 5,844,997, Dec. 1, 1998.

[24] M. Kompis, H. Pasterkamp, and G. Wodicka, Acoustic imaging of the human chest, *Chest*, **120**(4), 1309–1321 (2001). doi:10.1378/chest.120.4.1309

[25] Z. Moussavi, *Fundamentals of Respiratory Sounds and Analysis*, 1st edn, Morgan & Claypool, San Rafael, CA (2007). doi:10.2200/S00054ED1V01Y200609BME008

[26] Z. Moussavi, M.T. Leopando, H. Paterkamp, and G. Rempel, Computerized acoustical respiratory phase detection without airflow measurement, *Med. Biol. Eng. Comput.*, **38**(2), 198–203 (2000). doi:10.1007/BF02344776

[27] A. Yadollahi and Z. Moussavi, A robust method for estimating respiratory flow using tracheal sound entropy, *IEEE Trans. Biomed. Eng.*, **53**(4), 662–668 (2006). doi:10.1109/TBME.2006.870231

[28] L.J. Hadjileontiadis and S.M. Panas, A wavelet-based reduction of heart sound noise from lung sounds, *Int. J. Med. Inform.*, **52**, 183–190 (1998). doi:10.1109/IEMBS.2003.1279719

[29] I. Hossain and Z. Moussavi, An overview of heart-noise reduction of lung sound using wavelet transform based filter, *Proc. 25th IEEE Eng. Med. Biol. Soc. (EMBS)*, 458–461 (2003).

[30] V.K. Iyer, P.A. Ramamoorthy, H. Fan, and Y. Ploysongsang, Reduction of heart sounds from lung sounds by adaptive filtering, *IEEE Trans. Biomed. Eng.*, **33**(12), 1141–1148 (1986).

[31] L. Yip and Y.T. Zhang, Reduction of heart sounds from lung sound recordings by auto-mated gain control and adaptive filtering techniques, *Proc. 23rd IEEE Eng. Med. Biol. Soc. (EMBS)*, 2154–2156 (2001). doi:10.1109/IEMBS.2001.1017196

[32] L. Guangbin, C. Shaoqin, Z. Jingming, C. Jinzhi, and W. Shengju, The development of a portable breath sound analysis system, *Proc. 14th IEEE Eng. Med. Biol. Soc. (EMBS)*, 2582–2583 (1992).

[33] L. Yang-Sheng, L. Wen-Hui, and Q. Guang-Xia, Removal of the heart sound noise from the breath sound, *Proc. 10th IEEE Eng. Med. Biol. Soc. (EMBS)*, 175–176 (1988). doi:10.1109/IEMBS.1988.94464

[34] M. Kompis and E. Russi, Adaptive heart-noise reduction of lung sounds recorded by a single microphone, *Proc. 14th IEEE Eng. Med. Biol. Soc. (EMBS)*, 691–692 (1992). doi:10.1109/IEMBS.1992.595788

[35] J. Gnitecki, I. Hossain, Z. Moussavi, and H. Pasterkamp, Qualitative and quantitative eval-uation of heart sound reduction from lung sound recordings, *IEEE Trans. Biomed. Eng.*, **52**(10), 1788–1792 (2005). doi:10.1109/TBME.2005.855706

[36] M.T. Pourazad, Z. Moussavi, F. Farahmand, and R. Ward, Heart sounds separation from lung sounds using independent component analysis, *Proc. 27th IEEE Eng. Med. Biol. Soc. (EMBS)*, 2736–2739 (2005). doi:10.1109/IEMBS.2005.1617037

[37] M.T. Pourazad, Z. Moussavi, and G. Thomas, Heart sound cancellation from lung sound recording using adaptive threshold and 2D interpolation in time–frequency do-main, *Proc. 25th IEEE Eng. Med. Biol. Soc. (EMBS)*, 2586–2589 (2003). doi:10.1109/IEMBS.2003.1280444

[38] M.T. Pourazad, Z. Moussavi, and G. Thomas, Heart sound cancellation from lung sound recording using time–frequency filtering, *J. Med. Biol. Eng.*, **44**(3), 216–225 (2006). doi:10.1007/s11517-006-0030-8

[39] Z. Moussavi, D. Floras, and G. Thomas, Heart sound cancellation based on multiscale products and linear prediction, *Proc. 26th IEEE Eng. Med. Biol. Soc. (EMBS)*, 3840–3843 (2004). doi:10.1109/IEMBS.2004.1404075

[40] D. Floras, Z. Moussavi, and G. Thomas, Heart sound cancellation based on multiscale product and linear prediction, *IEEE Trans. Biomed. Eng.*, **54**(2), 234–243 (2007).

[41] C. Ahlstrom, O. Liljefeldt, P. Hult, and P. Ask, Heart sound cancellation from lung sound recordings using recurrence time statistics and nonlinear prediction, *IEEE Signal Process. Lett.*, **12**(12), 812–815 (2005).

[42] L.J. Hadjileontiadis and S.M. Panas, Adaptive reduction of heart sounds from lung sounds using fourth-order statistics, *IEEE Trans. Biomed. Eng.*, **44**(7), 642–648 (1997). doi:10.1109/10.594906

[43] R.L.H. Murphy, S.K. Holford, and W.C. Knowler, Visual lung sound characterization by time-expanded waveform analysis, *N. Engl. J. Med.*, **296**, 968–971 (1977).

[44] M. Ono, K. Arakawa, M. Mori, T. Sugimoto, and H. Harashima, Separation of fine crackles from vesicular sounds by a nonlinear digital filter, *IEEE Trans. Biomed. Eng.*, **36**(2), 286–291 (1989). doi:10.1109/10.16477

[45] L.J. Hadjileontiadis and S.M. Panas, Nonlinear separation of crackles and squawks from vesicular sounds using third-order statistics, *Proc. IEEE 18th EMBS Conf. (EMBS)*, **5**, 2217–2219 (1996). doi:10.1109/IEMBS.1996.646504

[46] L.J. Hadjileontiadis and S. M. Panas, Separation of discontinuous adventitious sounds from vesicular sounds using a wavelet-based filter, *IEEE Trans. Biomed. Eng.*, **44**(12), 1269–1281 (1997). doi:10.1109/10.649999

[47] Y.A. Tolias, L.J. Hadjileontiadis, and S.M. Panas, A fuzzy rule-based system for real-time separation of crackles from vesicular sounds, *Proc. 19th IEEE Eng. Med. Biol. Soc. (EMBS)*, 1115–1118 (1997). doi:10.1109/IEMBS.1997.756547

[48] P.M. Mastorocostas, Y.A. Tolias, J.B. Theocharis, L.J. Hadjileontiadis, and S.M. Panas, An orthogonal least squares-based fuzzy filter for real time analysis of lung sounds, *IEEE Trans. Biomed. Eng.*, **47**(9), 1165–1176 (1997).

[49] L.J. Hadjileontiadis and S.M. Panas, Enhanced separation of crackles and squawks from vesicular sounds using nonlinear filtering with third-order statistics, *J. Tenn. Acad. Sci.*, **73**(1–2), 47–52 (1998). doi:10.1109/IEMBS.1996.646504

[50] Y.A. Tolias, L.J. Hadjileontiadis, and S.M. Panas, Real-time separation of discontinuous adventitious sounds from vesicular sounds using a fuzzy rule-based filter, *IEEE Trans. Inf. Technol. Biomed.*, **2**(3), 204–215 (1998). doi:10.1109/4233.735786

[51] L.J. Hadjileontiadis, D.A. Patakas, N.J. Margaris, and S. M. Panas, Separation of crackles and squawks from vesicular sounds using a wavelet-based filtering technique, *COMPEL*, **17**(5/6), 649–657 (1998), MCB University Press. doi:10.1108/03321649810220973

[52] L.J. Hadjileontiadis, Y.A. Tolias, and S.M. Panas, Intelligent system modeling of bioacoustic signals using advanced signal processing techniques, in: C.T. Leondes (ed.), *Intelligent Systems: Technology and Applications*, Vol. 3, CRC Press, Boca Raton, FL, pp. 103–156 (2002).

[53] L.J. Hadjileontiadis and I.T. Rekanos, Detection of explosive lung and bowel sounds by means of fractal dimension, *IEEE Signal Process. Lett.*, **10**(10), 311–314 (2003). doi:10.1109/LSP.2003.817171

[54] L.J. Hadjileontiadis, Wavelet-based enhancement of lung and bowel sounds using fractal dimension thresholding—Part I: Methodology, *IEEE Trans. Biomed. Eng.*, **52**(6), 1143–1148 (2005). doi:10.1109/TBME.2005.846706

[55] L.J. Hadjileontiadis, Wavelet-based enhancement of lung and bowel sounds using fractal dimension thresholding—Part II: Application results, *IEEE Trans. Biomed. Eng.*, **52**(6), 1050–1064 (2005). doi:10.1109/TBME.2005.846717

[56] I.T. Rekanos and L.J. Hadjileontiadis, An iterative kurtosis-based technique for the detection of nonstationary bioacoustic signals, *Signal Process.*, **86**, 3787–3795 (2006). doi:10.1016/j.sigpro.2006.03.020

[57] L.J. Hadjileontiadis, Empirical mode decomposition and fractal dimension filter: a novel technique for denoising explosive lung sounds, *IEEE Eng. Med. Biol. Mag.*, **26**(1), 30–39 (2007), Art. No. 15.

[58] L.J. Hadjileontiadis and S.M. Panas, Autoregressive modeling of lung sounds using higher-order statistics: estimation of source and transmission, *Proc. IEEE Signal Processing Workshop on Higher-Order Statistics (SPW-HOS)*, 4–8 (1997). doi:10.1109/HOST.1997.613476

[59] L.J. Hadjileontiadis and S.M. Panas, Nonlinear analysis of musical lung sounds using the bicoherence index, *Proc. 19th IEEE Eng. Med. Biol. Soc. (EMBS)*, 1126–1129 (1997). doi:10.1109/IEMBS.1997.756551

[60] J. Gnitecki, Z. Moussavi, and H. Pasterkamp, Classification of lung sounds during bronchial provocation using waveform fractal dimensions, *Proc. 26th IEEE Eng. Med. Biol. Soc. (EMBS)*, 3844–3847 (2004). doi:10.1109/IEMBS.2004.1404076

[61] E. Conte, A. Vena, A. Federici, R. Giuliani, and J.P. Zbilut, A brief note on possible detection of physiological singularities in respiratory dynamics by recurrence quantification analysis of lung sounds, *J. Chaos Solitons Fractals*, **21**, 869–877 (2004). doi:10.1016/j.chaos.2003.12.098

[62] J. Gnitecki and Z. Moussavi, The fractality of lung sounds: a comparison of three waveform fractal dimension algorithms, *J. Chaos Solitons Fractals*, **26**(4), 1065–1072 (2005). doi:10.1016/j.chaos.2005.02.018

[63] J. Gnitecki, Z. Moussavi, and H. Pasterkamp, Geometrical and dynamical state space parameters of lung sounds, *5th Int. Workshop on Biosignal Interpretation (BSI)*, pp. 113–116 (2005).

[64] C. Ahlstrom, A. Johansson, P. Hult, and P. Ask, Chaotic dynamics of respiratory sounds, *J. Chaos Solitons Fractals*, **29**, 1054–1069 (2006). doi:10.1016/j.chaos.2005.08.197

[65] M. Mori, K. Kinoshita, H. Morinari, T. Shiraishi, S. Koike, and S. Murao, Waveform and spectral analysis of crackles, *Thorax*, **35**(11), 843–850 (1980).

[66] M. Matsuzaki, Polarity of crackle waveforms: a new index for crackle differentiation, *Hokkaido Igaku Zasshi*, **60**(1), 104–113 (1985).

[67] J. Hoevers and R.G. Loudon, Measuring crackles, *Chest*, **98**(5), 1240–1243 (1990).

[68] A. Cohen, Signal processing methods for upper airway and pulmonary dysfunction diagnosis, *IEEE Eng. Med. Biol. Mag.*, **9**(1), 72–75 (1990).

[69] M. Munakata, H. Ukita, I. Doi, Y. Ohtsuka, Y. Masaki, Y. Homma, and Y. Kawakami, Spectral and waveform characteristics of fine and coarse crackles, *Thorax*, **46**(9), 651–657 (1991).

[70] N. Al Jarad, S.W. Davies, R. Logan-Sinclair, and R. M. Rudd, Lung crackle characteristics in patients with asbestosis, asbestos-related pleural disease and left ventricular failure using a time-expanded waveform analysis—a comparative study, *Respir. Med.*, **88**(1), 37–46 (1994). doi:10.1016/0954-6111(94)90172-4

[71] N. Gavriely and D.W. Cugell, *Breath Sounds Methodology*, CRC Press, Boca Raton, FL (1995).

[72] A.R. Sovijarvi, P. Helistö, L.P. Malmberg, K. Kallio, E. Paajanen, A. Saarinen, P. Lipponen, S. Haltsonen, L. Pekkanen, P. Piirila, L. Naveri, and T. Katila, A new versatile PC-based lung sound analyzer with automatic crackle analysis (HeLSA); repeatability of spectral parameters and sound amplitude in healthy subjects, *Technol. Health Care*, **6**(1), 11–22 (1998).

[73] T. Kalayci, G. Celebi, Y. Ozturk, and M. Ozhan, Automatic detection and classification of crackles by using a neural network, *Proc. 14th IEEE Eng. Med. Biol. Soc. (EMBS)*, **6**, 2580–2581 (1992). doi:10.1109/IEMBS.1992.592865

[74] B. Sankur, Y.R. Kahya, E.C. Güler, and T.S. Engin, Comparison of AR-based algorithms for respiratory sounds classification respiratory disease diagnosis using lung sounds, *Comput. Biol. Med.*, **24**(1), 67–76 (1994). doi:10.1016/0010-4825(94)90038-8

[75] L. Pesu, P. Helistö, E. Ademovic, J.C. Pesquet, A. Saarinen, and A.R. Sovijarvi, Classification of respiratory sounds based on wavelet packet decomposition and learning vector quantization, *Technol. Health Care*, **6**(1), 65–74 (1998).

[76] L.J. Hadjileontiadis and S.M. Panas, On modeling impulsive bioacoustic signals with symmetric alpha-stable distributions: application in discontinuous adventitious lung sounds and explosive bowel sounds, *Proc. 20th IEEE Eng. Med. Biol. Soc. (EMBS)*, **1**, 13–16 (1998). doi:10.1109/IEMBS.1998.745810

[77] Y.P. Kahya and C.A. Yilmaz, Modeling of respiratory crackles, *Proc. 22nd IEEE Eng. Med. Biol. Soc. (EMBS)*, **1**, 632–634 (2000). doi:10.1109/IEMBS.2000.900823

[78] L.J. Hadjileontiadis, Discrimination analysis of discontinuous breath sounds using higher-order crossings, *Med. Biol. Eng. Comput.*, **41**(4), 445–455 (2003). doi:10.1007/BF02348088

[79] A. Kandaswamy, C. Sathish Kumar, R.P. Ramanathan, S. Jayaraman, and N. Malmurugan, Neural classification of lung sounds using wavelet coefficients, *Comput. Biol. Med.*, **34**, 523–537 (2004).

[80] C.A. Yilmaz and Y. P. Kahya, Modeling of pulmonary crackles using wavelet networks, *Proc. 27th IEEE Eng. Med. Biol. Soc. (EMBS)*, 7560–7563 (2005).

[81] C.A. Yilmaz and Y.P. Kahya, Multi-channel classification of respiratory sounds, *Proc. 28th IEEE Eng. Med. Biol. Soc. (EMBS)*, **1**, 2864–2867 (2006).

[82] O. Taketoshi, S. Hayaru, and K. Shoji, Discrimination of lung sounds using a statistics of waveform intervals, *IPSJ SIG Technical Reports*, **2006**(68(MPS-60)), 1–4 (2006).

[83] L.J. Hadjileontiadis, A texture-based classification of crackles and squawks using lacunarity, *IEEE Trans. Biomed. Eng.*, submitted (2008).

[84] J.M. Mendel, Tutorial on higher-order statistics (spectra) in signal processing and system theory: theoretical results and some applications, *Proc. IEEE*, **79**(3), 278–305 (1991). doi:10.1109/5.75086

[85] P.J. Huber, B. Kleiner, T. Gasser, and G. Dumermuth, Statistical methods for investigating phase relations in stationary stochastic processes, *IEEE Trans. Audio Electroacoust.*, **19**, 78–86 (1971). doi:10.1109/TAU.1971.1162163

[86] C.L. Nikias and A.P. Petropulu, *Higher-Order Spectra Analysis: A Nonlinear Signal Processing Framework*, Prentice-Hall, Englewood Cliffs, NJ (1993). doi:10.1109/IEMBS.1993.978564

[87] J.G. Proakis, C.M. Rader, F. Ling, and C.L. Nikias, *Advanced Digital Signal Processing*, Macmillan, New York (1992).

[88] S. Elgar and R.T. Guza, Statistics of bicoherence, *IEEE Trans. Acoust. Speech Signal. Process.*, **36**(10), 1667–1668 (1988). doi:10.1109/29.7555

[89] V.K. Iyer, P.A. Ramamoorthy, and Y. Ploysongsang, Autoregressive modeling of lung sounds: characterization of source and transmission, *IEEE Trans. Biomed. Eng.*, **36**(11), 1133–1137 (1989). doi:10.1109/10.40821

[90] M.R. Raghuveer and C.L. Nikias, Bispectrum estimation: a parametric approach, *IEEE Trans. Acoust., Speech Signal Process.*, **33**(4), 1213–1230 (1985). doi:10.1109/TASSP.1985.1164679

[91] C.L. Nikias and M.R. Raghuveer, Bispectrum estimation: a digital signal processing framework, *Proc. IEEE*, **75**(7), 869–891 (1987).

[92] A. Swami, J.M. Mendel, and C.L. Nikias, *Higher-Order Spectral Analysis Toolbox*, 3rd edn, The Mathworks, Natick, MA, chap. 1 (1998).

[93] L.J. Hadjileontiadis, Analysis and processing of lung sounds using higher-order statistics-spectra and wavelet transform, Ph.D. dissertation, Aristotle University of Thessaloniki, Thessaloniki, Greece, pp. 139–175 (1997).

[94] L.J. Hadjileontiadis and S.M. Panas, Higher-order statistics: a robust vehicle for diagnostic assessment and characterisation of lung sounds, *Technol. Health Care*, **5**(5), 359–374 (1997).

[95] C.L. Nikias and M. Shao, *Signal Processing with Alpha-Stable Distributions and Applications*, Wiley & Sons, Inc., New York, USA (1995).

[96]   W. Feller, *An introduction to probability theory and its applications*, Vol. II, Wiley & Sons, Inc., New York (1966).

[97]   J.H. McCulloch, Simple consistent estimators of stable distribution parameters, *Commun. Stat. Simul. Comput*, **15**(4), 1109–1136 (1986). doi:10.1080/03610918608812563

[98]   I.A. Koutrouvelis, An iterative procedure for the estimation of the parameters of stable laws, *Commun. Stat. Simul. Comput.*, **10**(1), 17–28 (1981). doi:10.1080/03610918108812189

[99]   M. Kanter and W.L. Steiger, Regression and autoregression with infinite variance, *Adv. Appl. Probab.*, **6**, 768–783 (1974). doi:10.2307/1426192

[100]   L.J. Hadjileontiadis, A.J. Giannakidis, and S.M. Panas, *a*-Stable modeling: a novel tool for classifying crackles and artifacts, in *Proc. 25th Int. Lung Sounds Association Conference (ILSA)*, H. Pasterkamp, ed., Chicago, IL (2000).

[101]   A. Grossmann and J. Morlet, Decomposition of Hardy functions into square integrable wavelets of constant shape, *SIAM J. Math. Anal*, **15**, 723–736 (1984). doi:10.1137/0515056

[102]   P.S. Addison, *The Illustrated Wavelet Transform Handbook: Introductory Theory and Applications in Science, Engineering, Medicine and Finance*, Institute of Physics (IOP) Publishing, Bristol, UK (2002).

[103]   N.M. Astaf'eva, Wavelet analysis: basic theory and some applications, *Physics-Uspekhi*, **39**(11), 1085–1108 (1996). doi:10.1070/PU1996v039n11ABEH000177

[104]   I. Daubechies, *Ten Lectures on Wavelets* (CBMS Lecture Notes Series), SIAM, Philadelphia, PA (1991).

[105]   M. Farge, Wavelet transforms and their applications to turbulence, *Annu. Rev. Fluid Mech.*, **24**, 395–457 (1992). doi:10.1146/annurev.fl.24.010192.002143

[106]   S.G. Mallat, *A Wavelet Tour of Signal Processing*, Academic Press, San Diego (1998).

[107]   S.G. Mallat, A theory for multiresolution signal decomposition: the wavelet representation, *IEEE Trans. Pattern Anal. Machine Intell.*, **11**(7), 674–693 (1989). doi:10.1109/34.192463

[108]   I. Daubechies, Orthonormal bases of compactly supported wavelets, *Commun. Pure Appl. Math.*, **41**, 909–996 (1988). doi:10.1002/cpa.3160410705

[109]   A. Cohen and J. Kovačević, Wavelets: the mathematical background, *Proc. IEEE*, **84**(4), 514–522 (1996). doi:10.1109/5.488697

[110]   M. Vetterli and J. Kovačević, *Wavelets and Subband Coding*, chap. 4, Prentice-Hall, Englewood Cliffs, NJ, pp. 201–298 (1995).

[111]   S.G. Mallat, Special issue on wavelets, *Proc. IEEE*, **84**(4), 507–686 (1996).

[112]   V. Gross, T. Penzel, L.J. Hadjileontiadis, U. Koehler, and C. Vogelmeier, Electronic auscultation based on wavelet transformation in clinical use, *Proc. 24th IEEE Eng. Med. Biol. Soc. (EMBS)*, pp. 1531–1532 (2002). doi:10.1109/IEMBS.2002.1106521

[113] V. Gross, T. Penzel, P. Fachinger, M. Fröhlich, J. Sulzer, and P. von Wichert, A simple method for detecting pneumonia with using wavelet-transformation, *Proc. Med. Biol. Eng. Comput. (EMBEC)*, **37**(Suppl. 2), 536–537 (1999).

[114] F. Ayari, A.T. Alouani, and M. Ksouri, Wavelets: an efficient tool for lung sounds analysis, in *Proc. IEEE Computer Systems and Applications (AICCSA)*, pp. 875–878 (2008).

[115] L. Ke and W. Houjun, A novel wavelet transform modulus maxima based method of measuring Lipschitz exponent, *Proc. International Conference Communications, Circuits and Systems (ICCCAS)*, pp. 628–632 (2007).

[116] S.G. Mallat and W. L. Hwang, Singularity detection and processing with wavelets, *IEEE Trans. Inf. Theory*, **38**(2), 617–643 (1992). doi:10.1109/18.119727

[117] A. Kandaswamy, C.S. Kumarb, R.P. Ramanathan, S. Jayaraman, and N. Malmurugan, Neural classification of lung sounds using wavelet coefficients, *Comput. Biol. Med.*, **34**, 523–537 (2004).

[118] M. Riedmiller and H. Braun, A direct adaptive method for faster backpropagation learning: the RPROP algorithm, in H. Ruspini (ed.), *Proc. IEEE Int. Conf. Neural Networks (ICNN)*, pp. 586–591 (1993). doi:10.1109/ICNN.1993.298623

[119] Y. Birkelund and A. Hanssen, Multitaper Estimators for Bispectra, *Proc. IEEE SP Workshop on Higher-Order (SPW-HOS)*, pp. 207–213 (1999). doi:10.1109/HOST.1999.778727

[120] B.Ph. van Milligen, E. Sánchez, T. Estrada, C. Hidalgo, B. Brañas, B. Carreras, and L. Garcia, Wavelet bicoherence: a new turbulence analysis tool, *Phys. Plasmas*, **2**(8), 3017–3032 (1995). doi:10.1063/1.871199

[121] T. Dudok de Wit and V.V. Krasnosel'skikh, Wavelet bicoherence analysis of strong plasma turbulence at the earth's quasiparallel bow shock, *Phys. Plasmas*, **2**(11), 4307–4311 (1995). doi:10.1063/1.870985

[122] Y. Larsen and A. Hanssen, Wavelet-polyspectra: analysis of non-stationary and non-Gaussian/non-linear signals, *Proc. IEEE Workshop on Statistical Signal and Array Processing (WSSAP)*, pp. 539–543 (2000). doi:10.1109/SSAP.2000.870183

[123] B.Ph. van Milligen, C. Hidalgo, and E. Sánchez, Nonlinear phenomena and intermittency in plasma turbulence, *Phys. Rev. Lett.*, **74**(3), 395–398 (1995). doi:10.1103/PhysRevLett.74.395

[124] Y.C. Kim and E.J. Powers, Digital bispectral analysis of self-excited fluctuation spectra, *Phys. Fluids*, **21**, 1452–1453 (1978). doi:10.1063/1.862365

[125] S.A. Taplidou and L.J. Hadjileontiadis, Nonlinear analysis of wheezes using wavelet bicoherence, *Comput. Biol. Med.*, **37**, 563–570 (2007). doi:10.1016/j.compbiomed.2006.08.007

[126] J.R. Duke Jr., J.T. Good Jr., L.D. Hudson, T.M. Hyers, M.D. Iseman, D.D. Mergenthaler, J.F. Murray, T.L. Petty, and D.R. Rollins, Frontline assessment of common pulmonary

presentations, in: J.F. Murray, L.D. Hudson, T.L., and Petty (eds.), *A Monograph for Primary Care Physicians*, Snowdrift Pulmonary Conference, Inc., Denver, CO, 2000. Available at: http://www.lungcancerfrontiers.org/pdf-books/asmnt_cmn_pulmryPrsntn.pdf.

[127] B. Kedem, *Time Series Analysis by Higher-Order Crossings*, IEEE Press, Piscataway, NJ (1994).

[128] L.J. Hadjileontiadis, Discrimination analysis of discontinuous breath sounds using higher-order crossings, *IEEE Med. Biol. Eng. Comput.*, **41**(4), 445–455 (2003). doi:10.1007/BF02348088

[129] N.E. Huang, Z. Shen, S.R. Long, M.L. Wu, H.H. Shih, Q. Zheng, N.C. Yen, C.C. Tung, and H.H. Liu, The empirical mode decomposition and Hilbert spectrum for nonlinear and nonstationary time series analysis, *Proc. R. Soc. Lond. A*, **454**(1971), 903–995 (1998).

[130] P. Gloersen and N.E. Huang, Comparison of interannual intrinsic modes in hemispheric sea ice covers and other geophysical parameters, *IEEE Trans. Geosci. Remote Sens.*, **41**(5), 1062–1074 (2003). doi:10.1109/TGRS.2003.811814

[131] N.E. Huang and S.S.P. Shen, *Hilbert–Huang Transform and Its Applications*, World Scientific, Singapore (2005).

[132] N.E. Huang and N.O. Attoh-Okine, *The Hilbert–Huang Transform in Engineering*, CRC Press (Taylor & Francis Group), Boca Raton, FL (2006).

[133] N.E. Huang, Z. Shen, and S.R. Long, A new view of nonlinear water waves—the Hilbert spectrum, *Annu. Rev. Fluid Mech.*, **31**, 417–457 (1999). doi:10.1146/annurev.fluid.31.1.417

[134] N.E. Huang, M.L. Wu, S.R. Long, S.S. Shen, W.D. Qu, P. Gloersen, and K.L. Fan, A confidence limit for the empirical mode decomposition and Hilbert spectral analysis, *Proc. R. Soc. Lond.*, **459A**, 2317–2345 (2003). doi:10.1098/rspa.2003.1123

[135] Z. Wu and N.E. Huang, A study of the characteristics of white noise using the empirical mode decomposition method, *Proc. R. Soc. Lond.*, **460A**, 1597–1611 (2004). doi:10.1098/rspa.2003.1221

[136] Z. Wu and N.E. Huang, Ensemble Empirical Mode Decomposition: a noise-assisted data analysis method, *COLA Technical Report 193*, 2005. Available at: ftp://grads.iges.org/pub/ctr/ctr_193.pdf.

[137] Z. Wu and N.E. Huang, Ensemble Empirical Mode Decomposition: a noise-assisted data analysis method, *Adv. Adapt. Data Anal.*, 2008 (in press). doi:10.1142/S1793536909000047

[138] J.R. Yeh, T.Y. Lin, J.S. Shieh, Y. Chen, N.E. Huang, Z. Wu, and C.K. Peng, Investigating complex patterns of blocked intestinal artery blood pressure signals by empirical mode decomposition and linguistic analysis, *J. Phys.: Conf. Ser. 96*, 1–7 (2008), doi:10.1088/1742-6596/96/1/012153. doi:10.1088/1742-6596/96/1/012153

[139]   K.T. Coughlin and K.K. Tung, 11-Year solar cycle in the stratosphere extracted by the empirical mode decomposition method, *Adv. Space Res.*, **34**(2), 323–329 (2004). doi:10.1016/j.asr.2003.02.045

[140]   S.C. Villalobos, R.G. Camarena, G.C. Lem, and T. A. Corrales, Crackle sounds analysis by empirical mode decomposition: nonlinear and nonstationary signal analysis for distinction of crackles in lung sounds, *IEEE Eng. Med. Biol. Mag.*, **26**(1), 40–47 (2007), Art. No. 15.

[141]   S.N. Rasband, Fractal dimension, in *Chaotic Dynamics of Nonlinear Systems*, , chap. 4, Wiley-Interscience, New York, pp. 71–83 (1997).

[142]   R. Esteller, G. Vachtsevanos, J. Echauz, T. Henry, P. Pennell, C. Epstein, R. Bakay, C. Bowen, and B. Litt, Fractal dimension characterizes seizure onset in epileptic patients, *Proc. IEEE Int. Conf. Acoust. Speech Signal Process. (ICASPP)*, **4**, 2343–2346, Phoenix, AZ (1999). doi:10.1109/IEMBS.1999.802520

[143]   W. Kinsner, Batch and real-time computation of a fractal dimension based on variance of a time series, *Technical Report, DEL94-6*, Dept of Electrical & Computer Eng, University of Manitoba, June 1994.

[144]   R.T.H. Laennec, *A Treatise on the Diseases of the Chest and on Mediate Auscultation*, 3rd edn. (J. Forbes, Trans.), Samuel Wood and Sons, and Collins and Hannay, New York, NY (1830).

[145]   B.B. Mandelbrot, *The Fractal Geometry of Nature*, Freeman, New York, NY (1983). doi:10.1119/1.13295

[146]   B. Lin and Z.R. Yang, A suggested lacunarity expression for Sierpinski carpets, *J. Phys. A*, **19**(2), L49–L52 (1986).

[147]   Y. Gefen, Y. Meir, B.B. Mandelbrot, and A. Aharony, Geometric implementation of hypercubic lattices with noninteger dimensionality by use of low lacunarity fractal lattices, *Phys. Rev. Lett.*, **50**(3), 145–148 (1983). doi:10.1103/PhysRevLett.50.145

[148]   C. Allain and M. Cloitre, Characterizing the lacunarity of random and deterministic fractal sets, *Phys. Rev. A*, **44**(6), 3552–3558 (1991). doi:10.1103/PhysRevA.44.3552

[149]   R.E. Plotnick, R.H. Gardner, W.W. Hargrove, K. Prestegaard, and M. Perlmutter, Lacunarity analysis: a general technique for the analysis of spatial patterns, *Phys. Rev. E*, **53**(5), 5461–5468 (1996). doi:10.1103/PhysRevE.53.5461

[150]   G. Du and T.S. Yeo, A novel lacunarity estimation method applied to SAR image segmentation, *IEEE Trans. Geosci. Remote Sens.*, **40**(12), 2687–2691 (2002).

[151]   R. Esteller, G. Vachtsevanos, J. Echauz, and B. Litt, A comparison of waveform fractal dimension algorithms, *IEEE Trans. Circ. Syst.*, **48**, 177–183 (2001). doi:10.1109/81.904882

[152]   P.S. Addison, *Fractals and Chaos: An Illustrated Course*, Taylor & Francis, Gloucester, UK (1997).

[153]   M. Katz, Fractals and the analysis of waveforms, *Comput. Biol. Med.*, **18**(3), 145–156 (1988). doi:10.1016/0010-4825(88)90041-8

[154]   C. Sevcik, A procedure to estimate the fractal dimension of waveforms, *Complexity International [Online]*, Vol. 5, 1998. Available at: http://journal-ci.csse.monash.edu.au/ci/vol05/sevcik/sevcik.html.

[155]   P. Dong, Test of a new lacunarity estimation method for image texture analysis, *Int. J. Remote Sens.*, **21**(17), 3369–3373 (2000). doi:10.1080/014311600750019985

[156]   W.J. Krzanowski, *Principles of Multivariate Analysis: A User's Perspective (Oxford Statistical Science Series)*, Oxford University Press, New York, NY, Rev Sub edition (2000).

[157]   G.A.F. Seber, *Multivariate Observations (Wiley Series in Probability and Statistics)*. John Wiley & Sons, Inc., San Francisco, CA (2004).

[158]   S.A. Taplidou, L.J. Hadjileontiadis, V. Gross, T. Penzel, and S.M. Panas, WED: an efficient wheezing-episode detector based on breath sounds spectrogram analysis, *Proc. 25th IEEE Eng. Med. Biol. Soc. (EMBS)*, pp. 2531–2534 (2003). doi:10.1109/IEMBS.2003.1280431

[159]   R. Loudon and R.L.H. Murphy, Jr., Lung sounds, *Am. Rev. Respir. Dis.*, **130**, 663–673 (1984).

[160]   R.R. Coifman and M.V. Wickerhauser, Adapted waveform 'denoising' for medical signals and images, *IEEE Eng. Med., Biol. Mag.*, **14**(5), 578–586 (1995). doi:10.1109/51.464774

[161]   P.J. Bickel and K.A. Doksum, *Mathematical Statistics*, Holden-Day, San Francisco, CA (1977).

[162]   G. Charbonneau, J.L. Racineux, M. Sudraud, and E. Tuchais, An accurate recording system and its use in breath sounds spectral analysis, *J. Appl. Physiol.*, **55**(4), 1120–1127 (1983).

[163]   B. Widrow, J.R. Glover, J.M. McCool, J. Kaunitz, C.S. Williams, R.H. Hearn, I.R. Zeidler, E. Dong, and R.C. Goodlin, Adaptive noise cancelling: principles and applications, *Proc. IEEE*, **63**(12), 1692–1716 (1975).

[164]   Z. Moussavi, R.M. Rangayyan, G.D. Bell, C.B. Frank, and K.O. Ladly, Screening and adaptive segmentation of vibroarthrographic signals, *IEEE Trans. Biomed. Eng.*, **43**(1), 15–23 (1996).

[165]   J.S.R. Jang, ANFIS: adaptive network-based fuzzy inference system, *IEEE Trans. Syst. Man Cybern.*, **23**(3), 665–685 (1993). doi:10.1109/21.256541

[166]   P.A. Mastorocostas and J.B. Theocharis, A recurrent fuzzy-neural model for dynamic system identification, *IEEE Trans. Syst. Man Cybern.—Part B*, **32**(2), 176–190 (2002). doi:10.1109/3477.990874

[167]   P.A. Mastorocostas and J.B. Theocharis, A recurrent fuzzy-neural filter for real-time separation of lung sounds, *Proc. IEEE Int. Joint Conf. Neural Networks (IJCNN)*, **5**, 3023–3028 (2005). doi:10.1109/IJCNN.2005.1556407

[168] P.A. Mastorocostas and J.B. Theocharis, A stable learning method for block-diagonal recurrent neural networks: application to the analysis of lung sounds, *IEEE Trans. Syst. Man Cybern.—Part B*, **36**, 242–254 (2006).

[169] S.C. Sivakumar, W. Robertson, and W.J. Phillips, On-line stabilization of block-diagonal recurrent neural networks, *IEEE Trans. Neural Networks*, **10**, 167–175 (1999). doi:10.1109/72.737503

[170] T.W. Lee, *Independent Component Analysis: Theory and Applications*, 2nd edn, chaps. 2 and 4, Kluwer, Dordrecht (2000). doi:10.1016/S0898-1221(00)00101-2

[171] J. Gnitecki and Z. Moussavi, Variance fractal dimension trajectory as a tool for heart sound localization in lung sounds recordings, *Proc. 25th IEEE Eng. Med. Biol. Soc. (EMBS)*, 2420–2423 (2003). doi:10.1109/IEMBS.2003.1280404

[172] F. Javed, P.A. Venkatachalam, and A.F. Hani, Knowledge based system with embedded intelligent heart sound analyser for diagnosing cardiovascular disorders, *J. Med. Eng. Technol.*, **31**(5), 341–350 (2007). doi:10.1080/03091900600887876

[173] L.A. Geddes, Birth of the stethoscope, *IEEE Eng. Med. Biol. Mag.*, **24**(1), 84–86 (2005).

Author Biography

Leontios J. Hadjileontiadis was born in Kastoria, Greece, in 1966. He received his B.S. degree in Electrical Engineering in 1989 and Ph.D. degree in Electrical and Computer Engineering in 1997, both from the Aristotle University of Thessaloniki, Thessaloniki, Greece. In December 1999, he began his stint as a faculty member in the Department of Electrical and Computer Engineering, AUTh, where he is currently an assistant professor, working on lung sounds, heart sounds, bowel sounds, electrocardiogram data compression, seismic data analysis, and crack detection in the Signal Processing and Biomedical Technology Unit of the university's Telecommunications Laboratory. His research interests are centered on higher-order statistics, α-stable distributions, higher-order zero crossings, wavelets, polyspectra, fractals, neurofuzzy modeling for medical, mobile, and digital signal processing applications. Dr. Hadjileontiadis is a member of the Technical Chamber of Greece, the IEEE, the Higher-Order Statistics Society, the International Lung Sounds Association (Secretary), and the American College of Chest Physicians. He won 2nd place in the Best Paper Competition of the 9th Panhellenic Medical Conference on Thorax Diseases '97, Thessaloniki. He was also an open finalist at the Student paper Competition (Whitaker Foundation) of the IEEE EMBS '97, Chicago, IL; a finalist at the Student Paper Competition (in memory of Dick Poortvliet) of the MEDICON '98, Lemesos, Cyprus; the recipient of the Young Scientist Award of the 24th International Lung Sounds Conference '99, Marburg, Germany; and the recipient of the 2nd Innovation Award, Research Committee, AUTh, 2008. In 2004, 2005, and 2007, he organized and served as a mentor to three 5-student teams that have won international awards at the Imagine Cup Competition (Microsoft) in Sao Paulo (2004), Yokohama (2005), and Seoul (2007), respectively, with projects relating to assistive technology for disabled people (blind, deaf, autistic kids). Dr. Hadjileontiadis also holds a Ph.D. degree in music composition (University of York, UK, 2004) and he is currently a Professor in composition at the State Conservatory of Thessaloniki, Greece.

Printed in the United States
by Baker & Taylor Publisher Services